To James

Best regards to a friend for many many years.

Ben

Energy—and How We Lost It

Energy—and How We Lost It

Johnny Mitchell

Pacesetter Press
A Division of Gulf Publishing Company
Houston, Texas

Energy—and How We Lost It

Library of Congress Catalog Card Number 78-062621
ISBN 0-88415-431-9

Editor's Foreword

It's not often that an editor gets to work with an author whose words of a decade or more ago turn out to be so incredibly prophetic. In sifting through Johnny Mitchell's personal and private papers to decide which would be apropos for a book such as this, time and time again I came across statements by Johnny concerning events which eventually came to pass, even though at the time he wrote them they were no more than a combination of his experience, intuition, and interpretation of events of the times.

A book such as this which chronicles in personal narrative an important segment of the history of a great industry and its interrelationship with the government of this country is an extremely ambitious undertaking, even for a man with the energy and tenacity of Johnny Mitchell. As his editor, however, it soon became apparent to me that to bury the accuracy of Johnny Mitchell's predictions within the text of his letters without spotlighting them for reader edification and reflection would be an editorial disservice. Therefore, I have convinced Johnny that we should excerpt some of his letters and put them up front. These excerpts will be of immediate benefit to historians and others concerned with putting history in its proper perspective as they can at any time delve deeper into the subject by reading the specific letters. The casual browser who will not read the letters, but will read some or all of these excerpts will also have benefitted. I take full responsibility for these excerpts since each is pulled out of context from the full letter.

Clayton A. Umbach, Jr.
Editor
Gulf Publishing Company
September, 1979

Dedication

I dedicate this book to two friends of many years—James A. Clark, a great columnist, a distinguished writer, and author, but best of all a critic who delighted in correcting me at all times.

I talked Jim into writing the introduction to this book. Jim left us the other day,—his desk was cluttered—with his own writing. . . .but he kept his promise—the introduction was completed.

Jim is probably looking down on me as I dedicate this book to him—his Irish eyes smiling as always—He has only been gone a short time and I already miss him.

To W. Earl Turner, economist, journalist, and author, who over these many years reviewed my letters and speeches, with one specific purpose—to temper my words—sometimes successful—sometimes not.

Finally, to my bride Alleyne who for so many years withstood the criticism from so many over my words and speeches.

Contents

Editor's Foreword v

Introduction
by James A. Clark* xii

		Excerpt on Page	Full Text on Page
From *The Washington Post*	July 24, 1958	1	—
Marlin Sandlin, *Vice President, Woodley Petroleum Company* to Johnny Mitchell	July 28, 1958	1	60
Johnny Mitchell to Marlin Sandlin, *Vice President, Woodley Petroleum Company*	August 8, 1958	2	61
Johnny Mitchell to General Ernest O. Thompson, Judge Olin Culberson, Mr. William J. Murray, Jr., *Texas Railroad Commission*	July 24, 1959	3	63
Johnny Mitchell to Ernest O. Thompson, Chairman, *Texas Railroad Commission*	July 27, 1959	3	—

*deceased

		Excerpt on Page	Full Text on Page
William J. Murray, Jr., *Commissioner, Texas Railroad Commission* to Johnny Mitchell	July 27, 1959	4	—
Olin Culberson, *Commissioner, Texas Railroad Commission* to Johnny Mitchell	July 27, 1959	5	—
Johnny Mitchell to Judge Olin Culberson, *Texas Railroad Commission* . .	August 7, 1959	5	64
From *Petroleum Week* . .	January 20, 1961	6	—
From *Petroleum Week* . .	March 31, 1961	7	—
Johnny Mitchell to Honorable Joseph C. Swidler, *Chairman, Federal Power Commission*	January 10, 1962	7	66
Johnny Mitchell to Honorable D. Otis Beasley, *Chairman, Oil Import Appeals Board*	March 16, 1962	8	68
Johnny Mitchell to Honorable Joseph C. Swidler, *Chairman, Federal Power Commission*	April 27, 1962	9	70
Johnny Mitchell to Honorable Cecil R. King, *U.S. Congressman*	May 2, 1962	9	71
Johnny Mitchell to General Ernest O. Thompson, *Chairman, Texas Railroad Commission*	June 10, 1962	10	—
Johnny Mitchell to the *Texas Delegation*	October 2, 1962	12	74
Johnny Mitchell to The Honorable Russell B. Long, *U.S. Senator*	October 13, 1962	12	76

		Excerpt on Page	Full Text on Page
Johnny Mitchell to Honorable Robert S. Kerr, *U.S. Senator*	September 17, 1962	11	73
Johnny Mitchell to Honorable Joseph C. Swidler, *Chairman, Federal Power Commission*	November 20, 1962	12	78
Johnny Mitchell to Honorable John Connally, *Governor of Texas*	February 3, 1963	13	79
Johnny Mitchell to Mr. F.J. Hortig, *State Land Commission*	March 6, 1963	—	82
Johnny Mitchell to The Honorable Alan Cranston, *Controller, State of California*	March 27, 1963	—	84
Johnny Mitchell to The Honorable Edmund G. Brown, *Governor of California*	March 29, 1963	—	85
Johnny Mitchell to Alan Cranston, *Chairman, State Lands Commission*	May 10, 1963	—	86
Johnny Mitchell to Alan Cranston, *Chairman, State Lands Commission*	May 30, 1963	14	87
Mr. R.G. Follis, Chairman of the Board, *Standard Oil Company of California*	July 9, 1963	15	90
Johnny Mitchell to Honorable Stewart L. Udall, *Secretary of the Interior*	July 10, 1963	15	92
John Connally, Governor, State of Texas to The Honorable Wilbur Mills, *Chairman, Ways and Means Committee*	July 11, 1963	16	94

		Excerpt on Page	Full Text on Page
Johnny Mitchell to The Honorable Clark Thompson, *U.S. Congressman*	July 24, 1963	17	96
Clark W. Thompson, *U.S. Congressman* to Johnny Mitchell	July 26, 1963	18	96
Johnny Mitchell to The Honorable John Connally, *Governor of Texas*	August 5, 1963	11	97
R.L. Foree to Johnny Mitchell	August 20, 1963	19	99
Johnny Mitchell to Honorable Dean Rusk, *Secretary of State*	November 7, 1963	20	100
Johnny Mitchell to Mr. Jack Rathbone, *Chairman, Standard Oil of New Jersey*	December 3, 1963	21	104
Johnny Mitchell to Mr. E. Bruce Street, *President, Texas Independent Producers & Royalty Owners Association* . . .	January 29, 1964	22	106
Johnny Mitchell to Mr. Ed J. Warren, *Chairman, American Petroleum Institute*	February 6, 1964	23	109
Johnny Mitchell to General Louis Pulaski . .	March 1, 1964	—	111
Johnny Mitchell to Mr. M.J. Rathbone, *Chairman, Standard Oil Company of New Jersey*	April 7, 1964	23	112
M.J. Rathbone, *Standard Oil Company of New Jersey* to Johnny Mitchell	June 10, 1964	24	114

		Excerpt on Page	Full Text on Page
Johnny Mitchell to Mr. Fred L. Hartley, *Union Oil Company of California*	June 10, 1964	25	115
Johnny Mitchell to Honorable Ernest O. Thompson, *Texas Railroad Commission*	July 23, 1964	26	117
Johnny Mitchell to Mr. John Swearinger, *Chairman, Standard Oil of Indiana*	August 16, 1964	27	120
Johnny Mitchell to Mr. Forrest N. Shumway, *President, Signal Oil and Gas Company*	September 9, 1964	28	122
Forrest N. Shumway, *President, Signal Oil and Gas Company* to Johnny Mitchell	September 22, 1964	29	124
Johnny Mitchell to Mr. Forrest N. Shumway, *President, Signal Oil and Gas Company*	October 9, 1964	29	125
Forrest N. Shumway, *President, Signal Oil and Gas Company* to Johnny Mitchell	November 4, 1964	30	128
Johnny Mitchell to Mr. Ralph W. Trueblood, Jr., *Vice President, Belridge Oil Company, President, Oil Producers Agency of California*	November 13, 1964	31	129
Johnny Mitchell to Honorable Nicholas deB. Katzenback, *Attorney General of the United States, Washington, D.C.*	June 9, 1965	31	134

		Excerpt on Page	Full Text on Page
Stanley Learned, *President, Phillips Petroleum Company* to Johnny Mitchell	July 15, 1965	32	137
Stanley Learned, *President, Phillips Petroleum Company* to The Honorable Nicholas deB. Katzenbach, *Attorney General of the United States, Washington, D.C.*	July 15, 1965	33	138
Robert L. Wright, *Acting Assistant Attorney General, Antitrust Division* to Johnny Mitchell	July 17, 1965	—	136
Johnny Mitchell to Mr. Stanley Learned, *President, Phillips Petroleum Company*	July 22, 1965	34	144
Johnny Mitchell to Honorable Nicholas deB. Katzenbach, *Attorney General of the United States, Washington, D.C.*	July 22, 1965	34	144
Honorable Russell B. Long, *U.S. Senator,* to Johnny Mitchell	August 27, 1965	—	147
Johnny Mitchell to The Honorable Russell B. Long, *U.S. Senator*	September 9, 1965	36	148
Jack Valenti, *Special Assistant to the President, The White House* to Johnny Mitchell	September 13, 1965	37	150
Johnny Mitchell to Jack Valenti, *Special Assistant to the President, The White House*	September 23, 1965	37	151

		Excerpt on Page	Full Text on Page
Olin E. Teague, *U.S. Congress* to Johnny Mitchell	September 30, 1965	37	152
Johnny Mitchell to Mr. Jake Hamon, *Independent*	October 15, 1965	38	154
Mr. John Doe, *Washington, D.C.*	December 1, 1965	39	157
Johnny Mitchell to The Honorable Olin E. Teague, *U.S. Congressman*	February 9, 1966	—	160
Johnny Mitchell to The Honorable Stewart L. Udall, *Secretary of the Interior*	May 1, 1966	—	161
Johnny Mitchell to Mr. Carl Reistle, *Chairman, Humble Oil & Refining Co.*	May 29, 1966	41	164
Johnny Mitchell to The Honorable Thomas H. Kuchel, *Washington, D.C.*	July 12, 1966	41	163
Johnny Mitchell to The Honorable Craig Hosmer, *Washington, D.C.*	October 12, 1966	42	168
Johnny Mitchell to Mr. R.C. McCurdy, *President, Shell Oil Company*	February 13, 1967	44	171
California Independent Producers & Royalty Association (CIPRO) to California Delegation, *United States Congress, Washington, D.C.*	February 27, 1967	45	173
Johnny Mitchell to Mr. John Doe, *Washington, D.C.*	May 15, 1967	45	174

		Excerpt on Page	Full Text on Page
Johnny Mitchell to Honorable Stewart Udall, *Secretary of the Interior* .	October 26, 1967	46	176
Johnny Mitchell to Honorable Ronald Reagan, *Governor of California* . .	July 8, 1968	47	177
Johnny Mitchell to Honorable Ronald Reagan, *Governor of California* . .	November 18, 1968	48	181
Johnny Mitchell to Mr. Houston Flournoy, *Chairman, State Land Commission, State of California*	November 20, 1968	49	184
Johnny Mitchell to Honorable Ronald Reagan, *Governor of California*	February 5, 1969	—	185
Johnny Mitchell to Mr. Michael L. Haider, *Chairman, Standard Oil Company of New Jersey* .	November 7, 1969	49	186
Michael L. Haider, *Standard Oil Company of New Jersey* to Johnny Mitchell	December 4, 1969	—	187
Johnny Mitchell to The Honorable Wilbur Mills, *U.S. Congressman*	December 2, 1969	50	188
Honorable George Bush, *U.S. Congressman* to Johnny Mitchell	December 8, 1969	—	189
Johnny Mitchell to The Honorable Olin E. Teague, *U.S.* Congressman	December 9, 1969	—	190
Johnny Mitchell to Mr. Clinton Engstrand, *Chairman, Liaison Committee of Cooperating Oil & Gas Associations* . .	May 7, 1970	50	190

		Excerpt on Page	Full Text on Page
Johnny Mitchell to Honorable Olin E. Teague, *U.S. Congressman*	June 16, 1970	—	192
Johnny Mitchell to Honorable Emilio Q. Daddario, *U.S. Congressman*	September 7, 1970	51	193
Johnny Mitchell to Mr. H.J. Haynes, *President, Standard Oil Company of California*	July 24, 1971	51	195
Johnny Mitchell to Mr. Peter Flanagan, *Executive Office of the President, Washington, D.C.*	February 10, 1972	52	198
Johnny Mitchell to Honorable George Wilson, *Executive Office of the President, Washington, D.C.*	June 10, 1972	53	200
Johnny Mitchell to Mr. D.F. McKeithan, Jr., *Chairman, Liaison Committee*	April 30, 1973	53	203
Johnny Mitchell to Honorable Henry M. Jackson, *U.S. Senator* . .	October 31, 1973	54	204
Johnny Mitchell to Mr. Robert O. Anderson, *Chairman, Atlantic Richfield Company*	January 2, 1974	55	206
Johnny Mitchell to Honorable Henry M. Jackson, *U.S. Senator* . .	April 23, 1974	56	209
Johnny Mitchell to The Honorable Bob Eckhardt, *U.S. Congressman*	June 27, 1965	57	212
Honorable Bob Eckhardt, *U.S. Congressman,* to Johnny Mitchell	August 14, 1975	—	214

		Excerpt on Page	Full Text on Page
Johnny Mitchell to The Honorable Frank Zarb, *Administrator, Federal Energy Administration, Washington, D.C.*	December 5, 1975	57	216
Johnny Mitchell to The Honorable Bob Eckhardt, *U.S. Congressman*	March 9, 1976	58	218
W.E. Turner, *Executive Vice President, TIPRO,* to Johnny Mitchell	May 10, 1978	—	221

Index 225

Introduction by James A. Clark*

There was a time in this country when there was too much oil and natural gas. This was a condition brought on by the curiosity, daring, freedom, and competitiveness of an individualistic breed of Americans know as wildcatters.

One of these was a happy-go-lucky son of Greek immigrants—Johnny Mitchell of Houston. He was one of a handful of wildcatters who saw the hand-writing on the wall a few years after the close of World War II and started making waves. His choice of weapons was speeches and letters. And he utilized his flare for the dramatic to fill the hall and banquet rooms for the speeches and newspapers the words he wrote, mostly to the chief executive officers of the giant integrated oil companies.

What happened after World War II to the oil and gas industry was an economic tragedy inflicted upon the people of the United States by a combination of ignorance, ambition, and deceit by national politicians from presidents down to the rookie congressmen and Senators, plus the avarice of the big companies that were the targets of Wildcatter Johnny Mitchell's unending volley of missives.

The economic tragedy came about as a result of two most unfortunate occurrences, both man-made. The first was a result of the discovery of oceans of oil in Venezuela and the Middle East, and then in Africa. How most of this fell into American hands is another story. The simple fact is that it did—into the hands of American major oil companies, especially the Famous Seven Sisters, five of them American.

These companies held off pouring imports into America from the Middle East for a while, although they started early shipping it from

*deceased

Venezuela, even before it was needed in the great war. At first they supplemented American supplies of oil, which brought on no objections from the independent wildcatters, and then they decided it was even better and more profitable to flood the American crude market, including their own, with oil which supplanted American production.

This action depressed the already low price of American crude and made life economically unbearable for thousands of independent explorers, and virtually so for those who were able to survive. Outside the individual wildcatters and their various organizations there was no weeping over this situation. The consumers reveled in the cheap prices and the majors grew richer on oil they bought for pennies in the Middle East and sold for dollars in the U.S., but below the living level of their "friends" the independents, in America.

Then, in 1954 came a decision by the United States Supreme Court which placed unrealistic price controls on natural gas and certified each producer, large or small, as a public utility. Again the gleeful, but short-sighted consumers, cheered and consumed their way directly toward a disaster which Mitchell and a few of his fellow independents and a few majors foresaw—an energy crisis.

Mitchell wrote nice letters to the big companies at first, and then scathing letters. He stirred up their entrails because copies of each of his letters went out to newspapers, magazines, congressmen, columnists, and radio stations, among others. And a copy of each letter went out to all of the other big company chief executives.

Johnny was not trained as a speaker or writer and he got little help from friends, most of whom were largely ignored because they, too, were immersed in not only a shortage of guts, but an ignorance born of not being oilmen themselves.

In the beginning, industry as a whole believed he was shooting from the hip in his warnings, but as time passed, the domestic oil industry began to take notice and listen. When they did it was all too late, the domestic oil industry was on its downhill slide.

James A. Clark

Excerpts

July 24, 1958

Excerpt

FROM: *The Washington Post*

"Senator Lyndon B. Johnson no longer represents Texas," Johnny Mitchell, a Houston independent oilman so stated in a gathering of oilmen discussing the reciprocal trade agreement act.

The controversy stems from a request from Texas independents through Mitchell as their spokesman to Senate Majority Leader Lyndon Johnson to press for moderate new guidelines into the trade expansion bill to curb excessive imports of foreign oil and it's products. Senator Johnson ignored Mitchell's request.

Mitchell concluded, "I don't know what Lyndon is doing up there in Washington . . . we want a man to be working for us and not be a national figure."

"We now have only one Senator representing Texas and like old horses in Kentucky that served their time we should bring Senator Johnson back home to Texas to pasture."

July 28, 1958

Excerpt of Letter

TO: Johnny Mitchell

FROM: Marlin E. Sandlin
Vice President
Woodley Petroleum Company

When I was in Washington recently, I was distressed to learn of your public criticism of Senator Lyndon B. Johnson regarding his efforts on behalf of the oil and gas industry. The newspaper article which I have before me indicates that you made the following statement, to-wit:

> "I don't know what Lyndon is doing up there in Washington—I want a man to be working for me and not to be a national figure."

It would be my hope that you would reconsider your attitude toward the Senator and that you would write him a letter correcting the record, not only for your own personal good but for the welfare of the entire oil and gas industry.

August 8, 1958

Excerpt of Letter

TO: Marlin Sandlin
Vice President
Woodley Petroleum Company

I deeply appreciate your letter admonishing me for my perhaps intemperate remarks at a public meeting about our able senior Senator.

In place of writing the Senator a letter, as you suggested, I will forward him a copy of this letter because it will be harder for me to repeat my feeling to him.

Marlin, I am sure you are aware that only a few oil men really devote a great deal of their time and effort to the oil and gas industry. These few, along with the various associations, have worked diligently to free our industry from FPC control and during the Congress of this year they worked especially hard to incorporate the ideas of the Ikard Amendment into the Reciprocal Trade Agreements Act. It had gained strength in the House, but the administration out-maneuvered the Congressmen in the House of Representatives and with the help of Sam Rayburn and Bob Anderson, this amendment was completely scuttled.

July 24, 1959

Excerpt of Letter

TO: General Ernest O. Thompson
Judge Olin Culberson
Mr. William J. Murray
Texas Railroad Commission

I was disturbed on reading the article written by one of our Houston independent operators on the Port Acres Hearing. I am sure that the three of you realize that I have sincere feelings on the question of wide spacing and am definitely opposed to drilling gas wells on one acre, or less, units. However, I would like you to know that I, in no way, agree with this operator and his attack upon your decision, as I feel that small acre drilling is strictly the fault of the producer and not the Commission. Even through I have expressed myself in this way to the industry, at no time to my knowledge have I ever felt personally opposed to the decisions handed down by your Commission.

The main point of this letter is that I shall continue to speak forth my thoughts on wide spacing and conservation but my public statements are intended against an *industry* which I feel needs correcting. I have a deep resentment toward producers who deliberately ruin the reservoir that nature has provided. It is my truthful opinion that I am personally fighting for a cause that has very little chance of success. However, even though I may not get industry support, the offenders, as far as I am concerned, should be publicly recognized.

July 27, 1959

TO: Mr. Johnny Mitchell

FROM: Ernest O. Thompson
Chairman
Railroad Commission of Texas

Dear Johnny:

Rule 37 is the spacing rule. The small tracts require an exception to the Rule 37.

It is the law that requires us to grant permits on small tracts. The courts have held we must give permits for at least one well on separate tracts not subdivided after oil and gas development.

It is not the examiners that prescribe the 2/3 acreage and 1/3 well formula, it is the Commission that makes this decision.

Sincerely,

Ernest O. Thompson
Chairman

July 27, 1959

TO: Mr. Johnny Mitchell

FROM: William J. Murray, Jr.
Commissioner
Railroad Commission of Texas

Dear Johnny:

I have just received and carefully read your letter of July 24th addressed to the three Commissioners. I think that it is splendidly written and I feel that all three Commissioners will appreciate receiving it. Since you already know pretty well how I feel on these matters, there is no point in further elaborating but I did want to let you know how well I think you have handled the situation.

Jo continues to improve, is gradually getting her strength back, and has had no complications from the operation. She has had no severe attack of the chest pain since the operation but does continue at times to have minor pain in the same spot. Doctor Bailey, the surgeon here, told Jo he had received a very nice letter from Ed White.

Again, may I say thanks to all of you for your sympathy, helpfulness, and understanding during our period of anxiety.

Kindest personal regards,

William J. Murray, Jr.

July 27, 1959

TO: Mr. Johnny Mitchell

FROM: Olin Culberson
Commissioner
Railroad Commission of Texas

Dear Johnny:

Yours of the 24th. Tell your critical friend to read the most recent decision of Halbouty-vs-Darsey and the Commission, which the 3rd Court of Appeals handed down ten days ago.

There are some people who do not seem to realize that this is still a government of laws and not of men. Criticism of my official actions never disturbs me. Even when those criticisms are the result of inconsistency.

I thoroughly agree with your statement that you and others are fighting a losing fight on wide spacing and you will never win it, regardless of the fact that the majors are behind it and using you boys who *used to be* independent until your interests began to run parallel with that of the majors, until the legal concept of property rights in Texas is changed.

I do not know of any one from whom I had rather take a licking on this subject than you, my friend, but "I ain't gonna take one."

Sincerely,

Olin Culberson

August 7, 1959

Excerpt of Letter

TO: Judge Olin Culberson
Texas Railroad Commission

Even though I consider myself a staunch friend and an admirer of yours, I am taking a decided issue on a subject that I believe you are dead wrong on. The subject deals with drilling wells on small tracts and your continuing determined stand backing the supposed "Little Fellows" who drill on these small tracts.

Great injustices are permitted to be committed by wealthy concerns which are using what you term the "Little Fellows" as a front man.

The exception to Rule 37 has created a hunting license for wealthy groups to hunt for small tracts in proven producing reservoirs. These wealthy groups under the guise of being small tract drillers begin their raid on regulated oil and gas reservoirs. These small tracts receive oil and gas allowable that unfairly drains what gas and oil that lies under their small tract in one day's normal allowable schedule, yet they receive an allowable for the life of the well. The "Law of Capture" is making this group richer and ruining reservoirs. The Commission is giving them a "license to steal."

January 20, 1961

Excerpt

FROM: *Petroleum Week*

"It is the Commission's job to regulate the industry, not to serve it."

This statement by Johnny Mitchell, Houston, President of Jade Oil Co., in a letter to stockholders, clearly states the duties of the Texas Railroad Commission.

Earlier in the letter, he had said:

> We must remember that the Texas Railroad Commission is not an industrial Commission but is a public regulatory body, designed not to serve the industry, but to serve the public . . .

"It is the people who are beginning to get stirred up about conservation because, after all, we're talking about the conservation of natural resources.

"If, however, the independents meet and agree they want changes in our statutes (relative to well spacing, unitization, and economy), then it follows that the commission will give the fullest consideration to the requested changes consistent with the needs of the people, whom they serve . . .

"The fault of conservation does not lie so much with the commission as with independent oil men who have not bothered to study their own problems."

March 31, 1961

Excerpt

FROM: *Petroleum Week*

The petroleum industry will become as exact as an accounting sheet.

That, in a few words, sums up the ultimate result of the Texas Supreme Court's opinion in the Normanna field case.

There can be no other conclusion reached from the opinion, which has been heralded and attacked within the industry, unless it is upset by legislative action or by the court itself.

The consensus of lawyers and others closely concerned with the issue is that the opinion overturns the "rule of capture."

The Texas Railroad Commission, because of a previous opinion from the Texas court, has operated under that rule in establishing proration formulas. The previous edict not only said a person was entitled to drill his property, however small, but also at a reasonable profit.

With the rule of capture overturned, the person who drills a small tract will be entitled to only the oil or gas which underlies his property.

And the burden of proof of establishing such reserves will fall upon the operator. Of course, he'll have assistance from offset operators.

January 10, 1962

Excerpt of Letter

TO: Honorable Joseph E. Swidler
Chairman
Federal Power Commission

I am fulfilling my promise to you, explaining in more detail what is happening in the interstate and intrastate gas markets here in Texas.

The situation has saturated the interstate gas market with "beefed up" reserves with "pay or take" contracts, with "buffer intrastate" lines producing gas on State of Texas regulations which are not only old and chaotic but a legal permit for a "license to steal."

The time has come for the Federal Power Commission, interstate gas pipeline companies, and independent gas producers to meet and explore the serious problems in the inter and intrastate contractual sales of natural gas. Variation in contractual arrangements allowed under current regulations may very well endanger normal growth in gas reserves if something is not done soon. The situation is an explosive one, and independent producers are in no mood to tolerate a continuation of some unethical and perhaps illegal practices.

March 16, 1962

Excerpt of Letter

TO: Honorable D. Otis Beasley
Chairman
Oil Import Appeals Board

I have noticed in the Wednesday edition of The Oil Daily that Texaco, Inc. has requested an additional oil import quota, stating the present crude oil stock situation on the West Coast is a potential threat to the national security.

Since no one has apparently challenged the Texaco statement, and since our company is a producer of California crude, I would like to use this means to protest the Texaco representation to your Board.

Texaco has been termed a "lone wolf" in the crude oil purchasing business, steadfastly refusing to engage in many, if any, of the normal exchanges of crude oil with other companies. For example, the Texas Railroad Commission has, on numerous occasions, asked Texaco representatives if they would purchase or exchange oil with other purchasing companies if the Railroad Commission set the allowable at a figure lower than Texaco, unlike other companies represented in those hearings will never promise to buy a single barrel of crude oil from other purchasing companies in order to balance their undersupply with the other companies' oversupply.

April 27, 1962

Excerpt of Letter

TO: Honorable Joseph C. Swidler
Chairman
Federal Power Commission

I am taking the liberty to bring up again the same serious problem, concerning recent developments in the F.P.C.'s involvement in issuing certificates to interstate pipeline originating in Texas, whose reserves are questionable.

It is my understanding that some of the long lines have matters now pending before your commission where careful look into the ratio of reserves in place as compared to the actual withdrawals of gas might appropriately be taken by the commission.

A close examination of such subsidiary operations and their interstate lines, the actual reserves dedicated to each such subsidiary line, and the volumes of deliveries made by and through such lines, would reveal, "beefed up" reserves and other facts not dreamed of by you as Chairman but certainly should be aware of by your examiner.

You, as Chairman, would also be surprised to find that numerous corporations with small gathering systems, have been organized and are being operated ostensibly as intrastate lines. When in fact they were always intended for and are engaged almost exclusively in interstate business.

May 2, 1962

Excerpt of Letter

TO: Honorable Cecil R. King
U.S. Congressman

As an independent oil producer in California, I feel it is not out of order for me to attempt to let you know how urgently we feel we need some Congressional mandate concerning the limitations on oil imports.

Whatever the need for liberalizing our foreign trade policy in accordance with the President's Trade Expansion Bill (H.R. 9900), I believe you will agree that certain unique situations require individual attention. You are aware that one of these is petroleum, and that the maintenance of a healthy home oil producing industry is absolutely essential to our nation's security. I believe that the home oil producing industry simply cannot survive intact unless proper restraints are placed on oil imports.

I have no sympathy for any domestic industry which wants protection simply to further the economic objectives of individuals. The domestic producers I know do not advocate any drastic change in the import pattern. We simply want to share in this nation's oil market to the extent that we can hold together a healthy producing segment adequate to meet emergencies.

We very much feel that unless the national security clause of the proposed Trade Expansion Bill is amended to provide some more positive relationship between imports and domestic production of this defense-vital commodity, the pressures will grow too great to contain imports. The Executive Department has enjoyed ample authority to limit imports since the 1955 extension of the Trade Agreement Act, but has simply proved unable to do so because of the enormous pressures from the State Department and the importing companies.

June 10, 1962

Excerpt of Letter

TO: General Ernest O. Thompson
Chairman
Texas Railroad Commission

I have always looked up to your leadership as the champion of our state's conservation system which accounts for the respect this nation has for you and our Texas Railroad Commission.

Conservation of our natural gas resources prompts me in writing you today. You have heard my story before but I cannot get off of the subject . . . "Ratable take proration for natural gas." The absolute necessity for it under the present conditions, the fact that it is in the public interest and the short sightedness of the forces that are opposing it.

Ratable take proration of gas is the only means of conserving this state's vital gas resources. It is also the only means of preserving equity among producers and the only means of assuring that all elements of the gas industry survive.

The solution you've told me before is difficult, but General, it must be done. Our gas proration system set by our outdated statutes is at best so confusing that its effectiveness is obscure. There is at its best a flagrant discrimination in the quantities taken between fields in an area and even between wells in the same field. Independents who have drilled their wells for maximum recovery find it hard to get a market for their gas wells in the field from which a gas pipeline system is drawing from its own wells or "special privileged wells" they have contracted with.

September 17, 1962

Excerpt of Letter

TO: Honorable Robert S. Kerr
U.S. Senate

At one time in recent weeks the independent oil producers' hopes were buoyed upward by your assurance that you planned to press for moderate new guidelines in the reciprocal trade agreement act to curb excessive imports of foreign oil and their products.

Our hopes were ill founded when we learned from reliable sources that you had no intention of going through with your proposed plan to press for these guidelines in the Senate.

I understand the votes were there in the Senate to approve the amendment on the floor of the Senate. Your leadership could have established a stable ratio between total imports of oil products and domestic production.

I am aware that your own company enjoys a sizeable daily import quota along with the big boys. These quotas are worth thousands of dollars a day to your company. Yet on the other hand you are an elected Senator from the State of Oklahoma where hundreds of domestic producers were looking toward you with the same trust as every independent in America.

This won't happen in your next few years in the Senate but there will come a day when you wished you would have pressed for moderate guidelines on oil imports. That day will come when unreliable imports of foreign oil becomes a way of life for every American.

October 2, 1962

TO: The Texas Delegation

When the roof falls in, when the Arabs take control of the world, when America becomes totally dependent on Middle East oil, who will take the blame; not Presidents that time will see out of office, not Secretary of Interiors, like Udall, who we pray will go out with President Kennedy; but the blame will be with you gentlemen who will stay with the ship of state. Most of you will probably be in office when the roof does fall in and the camel drivers from the Middle East become their self appointed dictators of the free world. It will then be too late.

This is why I am addressing this letter to you gentlemen in the Texas Delegation.

October 13, 1962

TO: The Honorable Russell B. Long
United States Senate

Senator, before too many years, imports will be exceeding our domestic production. We've lost the oil wildcatters who kept our nation strong for so many years. Our reserve producing capacity is no longer ours to fall back on. On any given day, at the slightest provocation, the Middle East will close down their wells and issue their proclamation of world supremacy. If you and your colleagues don't believe this is possible, brace yourself, 'cause it's coming.

November 20, 1962

Excerpt of Letter

TO: Honorable Joseph C. Swidler
Chairman
Federal Power Commission

We take this informal means of respectfully suggesting that you personally invite our detailed comments concerning recent developments in the FPC evolvement of an area pricing concept for natural gas production. This unusual procedure is warranted, we believe, both because (1) of assurances you gave at our annual meeting last May that interests of nonintegrated producers would not be overlooked in the development of an FPC area price concept, and (2) of indications that the industry advisory committee you have named may be unduly dominated by interests contrary to our own.

In particular, we would be pleased to detail for your personal attention our differences with findings and proposals of the FPC staff as stated at the re-opening of Permian Basin area rate proceedings November 15. We doubt that any worthwhile purpose will be served, however, unless we are given by you some reason to believe that our views will not be ignored or relegated to the staff for interpretation and possible distortion.

It is not unreasonable to ask, in view of recent trends, whether the FPC staff predominantly regards area pricing as an opportunity for the Commission to pose as champion of the consumer while in fact promulgating policies tending to promote monopoly practices—which in turn guarantee a kind of federal regulation that is in fact contrary to the interests of consumers and the nation as a whole.

May 2, 1963

Excerpt of Letter

TO: Honorable John Connally
Governor of Texas

As an admirer of yours, I find it somewhat difficult to tell you what you already know. That our State of Texas, which has assumed leadership among all the producing states in the matter of oil and gas conservation, has done nothing to prevent waste and protect the correlative rights where natural gas production is concerned.

Our present day concept of "state conservation" leans heavily on market demand proration." Prorating oil to market demand, despite suspicions of it from within and outside the industry since its inception, has in the main served the best interests of producers and consumers alike. In addition, it has provided a means for developing reserve-productive capacity which proved so essential to our security on several notable occasions. It is soundly based in "physical conservation," but primarily serves as a means for "protecting correlative rights." You recall that our Texas conservation statutes equate "the protection of correlative rights" with the "prevention of physical waste," and market demand proration serves both these objectives.

Our gas proration system in Texas is largely a hodge-podge of contradictory and confusing regulations which obscure their real ineffectiveness. In a very real sense they total up to the same kind of chaotic situation which in the East Texas situation of the 30's was considered a "conservation crisis." Wellhead gas prices range without logical basis from 3 or 4 cents to 30 cents per thousand, with wide price differentials occurring in the same field. Even more important, and certainly more directly related to conservation, there is flagrant discrimination in the quantities taken both as between fields in an area and even between wells in the same field. Independents often find it hard to get a market for their gas wells in a field from which a pipeline is drawing from it's own or other wells.

May 30, 1963

Excerpt of Letter

TO: Alan Cranston
Chairman
State Lands Commission of California

What benefit will be gained by Governor Brown, the State Lands Commissioners and O'Sullivan by the State's acceptance of the Chapman report? Political influence so obvious in the Chapman case is sure to hurt the Governor, O'Sullivan and the State Lands Commission. No political reward is worth this sacrifice. No visit to Pauley's Hawaiian Island could be worth it, nor is a trip to his fishing retreat in Mexico worth it.

Where Governor Brown, Pauley and O'Sullivan, and the State Lands Commission have welcomed Mr. Chapman with open arms and are paying him a fee of $35,000.00, you can be assured that I feel that any man who fought against state's rights as Chapman has in the past, should be about as welcome as the boll weevil and the fruit fly.

July 9, 1963

Excerpt of Letter

TO: Mr. R.G. Follis
Chairman of the Board
Standard Oil Company of California

Up to now there has been no hope of industry statesmanship among our many major producers. However, you too have fought the wars of major competition and you have helped build one of the great companies of our era.

Secondly, and most important, your company has grown in a state that has seen your fellow independent producers rapidly going out of existence. What greater contribution could you make to the industry than to ensure the survival of the California independents. A little less profit on a barrel of California low gravity crude could be your company's contribution to this industry. A few cents a barrel increase in the price of the small independents' crude will automatically mean their survival.

Perhaps the request is too much to ask of one man and one company to accept. But, I firmly believe that if I had the fortune to be the head of a company as large as Standard of California and had my roots embedded as firmly as your company has in this state, I would devote some of my company's energies to the welfare of this domestic industry.

If I had my choice of walking with kings who control foreign oil, or walking amongst the battered derricks, the leaking tanks, and the worn out equipment of the hard pressed independents of California; I would forego the privilege of walking with kings, until my fellow producers were able to walk alone.

July 10, 1963

Excerpt of Letter

TO: Honorable Stewart L. Udall
Secretary of the Interior

I feel impelled to express grave personal reservations regarding the "technical change" in the oil import program announced June 11.

My objection is not alone to the fact that this change cancelled out a relatively small reduction in crude oil imports east of the Rockies for the six-months' period beginning July 1. This scheduled nominal reduction did have special meaning, however, to the hard-pressed domestic oil producing industry—largely because it would have provided a long-awaited symbol of "stability" which the home producing industry badly needs to rebuild its confidence.

I feel quite certain that you could not have been aware that changing the program in this particular way would directly violate what were believed to be specific assurances from this Administration. Assurances on this precise point, as a matter of fact, were understood to have been given by White House Adviser Myer Feldman, Assistant Secretary John Kelly, Congressman Wilbur Mills, the late Senator Robert Kerr, the Senator Russell Long, all of whom indicated they were reflecting Presidential concurrence.

July 11, 1963

Excerpt of Letter

TO: The Honorable Wilbur Mills
Chairman
Ways and Means Committee

FROM: John Connally
Governor
State of Texas

As Governor of Texas, I wish to express in the strongest possible terms my grave concern over the committee's tentative adoption of tax changes adversely affecting the health of this State's leading industry.

In particular, I refer to the proposal that gain from the sale of a mineral property be taxed as ordinary income to the extent that the taxpayer has taken deductions for intangible drilling costs. In my letter of March 23, addressed to you as Chairman of the Ways and

Means Committee, I stressed particular objections to this proposal in these words:

"This proposal would have a particularly heavy impact *on the small operator* who is attempting to expand and grow by searching for and developing of vital new reserves, then selling them to accumulate dollars to reinvest in new properties with a view to finding and developing new oil. This is a common practice in the oil industry in Texas, a practice which should be encouraged not hampered. Secretary Dillon has testified that this 'recapture' provision would bring in an additional 50 million dollars each year. This burden would fall *on the small producer* because the larger units in the oil industry do not normally sell properties, but rather purchase them. This proposal would reduce selling and reinvestment activities and thus discourage wildcatting for new reserves."

July 24, 1963

Excerpt of Letter

TO: The Honorable Clark Thompson
U.S. Congressman

I have returned to California where I will stay until late August. I wish to express to you my sincere appreciation for not only your time, but your sincere efforts and understanding of the problem facing the independents at this present time.

For the first time, I left Washington feeling that through you and your efforts I was able to leave a small image of what we are facing in our fight to survive in this complex international oil industry. Naturally, we independents have the courage to keep fighting and with friends such as yourself, I sincerely feel not that we may, but we will survive these crises.

My very best to you and a thousand thanks from me and this appreciation, I can assure you, comes from the many independents that you haven't had the pleasure of meeting.

July 26, 1963

Excerpt of Letter

TO: Johnny Mitchell

FROM: Clark W. Thompson
U.S. Congressman

Thanks for your letter of the twenty-fourth.

The most effective job in behalf of the independents was done by your good self in Wilbur Mills' office. I did not realize how thoroughly you had sold him until I heard him outline to the Committee the typical example which you had given him of the way in which you finance your various projects. Certainly he was very much in accord with your thinking.

We have not taken final action and, of course, anything might happen but I want you to know that as I write this I am very optimistic. Whenever we do act I will get in touch with you immediately.

August 5, 1963

Excerpt of Letter

TO: The Honorable John Connally
Governor of the State of Texas

I have a copy of your letter addressed to Congressman Wilbur Mills dated July 11, 1963, concerning the intangible feature of the present proposed tax bill. Your concern that our intangibles would be taxed as ordinary income equal to our intangible deduction at the time of sale of these minerals prompts me to write.

I have been asked by our mutual friend, Robert L. Foree, Sr. and the executive committee to advise you of my experience with Congressman Wilbur Mills in a meeting we had in his chamber with Congressman Clark Thompson who arranged the meeting. Included in the meeting was Robert L. Foree, Sr. of Dallas and Earl Turner of Austin.

In the first few minutes of our conversation, I realized that Congressman Mills was really unaware of the significance of this provision to the small producer, and for the next hour I explained to him what it meant to the small producers and their investors. I advised him that the small producer was dependent on the small investor for capital for their risk ventures. I asked him if he objected. He replied that he thought the oil industry carried great risks and great responsibilities and that an oilman was entitled to the intangibles and depletion. I proceeded to explain to the Congressman, and he took notes on a scratch pad. Immediately he realized that no independent or investor could afford to drill, take his intangible costs as deductions the year of drilling, sell at a later date and then have the same intangibles become ordinary income at the time of such sale.

The Congressman looked at his colleague, Congressman Clark Thompson, who was sitting in on our meeting, and said the Committee had made a grave mistake. They were led to believe that this new proposed feature would be the least harmful to the independent oilman. I suggested that his information must have been furnished by major oil companies who never sell and who would never be affected by this provision. There was no denial by either Congressman, and I did not press the issue.

The dramatic part of this whole episode, as related later by Congressman Thompson, was watching Congressman Mills get up before the Ways and Means Committee, take out the crumpled scratch paper that he had made notations on during our conversation, and advise the Committee that he proposed that this provision be deleted from the tax bill. As you know, the provision was deleted.

August 20, 1963

Excerpt of Letter

TO: Johnny Mitchell

FROM: R.L. Foree

I want to take this method of trying to say "thank you a million" for your kind and wonderful letter of August 10, 1963, and the previous communications that you have sent me during the year 1962. You have been most generous in your kind words and thoughts. I only hope that I did assist you in making the great success you made as President of TIPRO.

Your active, dedicated, enthusiastic and unselfish efforts continued TIPRO on its ever upward service to the independent oil producers of the United States of America.

Your presentation of actual facts concerning the domestic oil industry to Chairman Wilbur Mills of the Ways and Means Committee of the U.S. House of Representatives was the main act that helped steer the ship of state away from damaging reefs and rocks and in the direction of equalization of domestic producers with international ones. You were in the right place at the right time with the right set of facts that enabled the Branch of the Federal Government regarding the tax bill, and had it not been for you, I feel positive that the domestic oil industry would have sunk lower, and the damage that would have been done, might have been irreparable.

November 7, 1963

Excerpt of Letter

TO: Honorable Dean Rusk
Secretary of State

It has become obvious that American diplomacy abroad has become subservient to the interest of international oil companies.

Now I'm for American exports, just as I'm for unsubsidized foreign investments by American companies—investments which are actually made to fill a need with risk capital of those companies. But when, as now, in the case of oil, these companies are investing abroad only because there's American subsidies to be gained, and when they're producing oil in excess of Foreign demand and increasing their imports of that excess oil into the U.S., weakening the price of crude worldwide—then I lose my enthusiasm for that kind of internationalism. I say even the companies involved, if they would look up a minute from their selfish pursuit of indirect subsidies, would realize that in the end they're hurting themselves not to mention their own country and the free world.

Now I admit I'm prejudiced in favor of American interests anywhere in the world, and I don't like to see American companies kicked around by any high-handed, dictatorial, socialistic despot. But neither am I too happy to see a few selfish companies sitting around expecting to have their chestnuts pulled out of fires they had no business starting. If American companies wish to invest their own capital to develop oil in any country in the world, and are willing to stand the considerable risk of expropriation, then I'm for them. But when they try to take advantage of a U.S. foreign aid situation to develop oil which is not needed in this period of surplus—oil which would not be developed if it were not in part underwritten by our foreign aid and subsidized by our tax favoritism, then I just can't get too concerned over them.

December 3, 1963

Excerpt of Letter

TO: Mr. Jack Rathbone
Chairman
Standard Oil Company of New Jersey

I hope you won't object to an old friend questioning a recent statement made by Mr. Emilio Collido, a former undersecretary of the U.S. State Department and now a Standard Vice-President; defending your Standard's increasing expenditure in developing oil in foreign areas.

He states: "Standard of New Jersey's export of capital equipment in the development of foreign oil far exceeded our payments for imported oil. Standard was helping our balance of payment."

His statement could ring true only if he closed his eyes to the fact that our nation was now being flooded with unreliable imports of oil. I am sure he didn't check with the present "total" import level of oil and products. For his information, it has reached an alarming 29 percent, and imports are on the increase. When Congress first wrote the National Security Clause into the Trade Bill in 1955, total imports bore a ratio to domestic production of 16.6 percent. The intention was to stabilize imports at that level. What are they today? And what of tomorrow? It's time to get concerned.

Mr. Collido continues: "Foreign development by Standard and other companies is an important program that would ensure our nation having dependable oil in the future." He knows better, the assumption that we can depend on Middle East, Venezuela and other foreign sources for a reliable source of oil for our future is only for a fool to hang his hat on.

The producing nations of the Middle East have experience abuse from our State Department and complete domination by our major international oil companies. This domination and abuse will not go on forever until their total oil wealth is dissipated feeding the oil hungry free world. Their day has to come.

The day will come when all the member nations band together in spite of their border differences, turn the valves of every oil well to the right and watch the oil hungry free world including our invincible, resourceful United States, fall to its knees, subdued and begging.

I don't know whether you and I will be around when this happens, but the vast resources of energy in the Middle East is destined to play a major role in the everyday life of every nation of the free world.

January 29, 1964

Excerpt of Letter

TO: Mr. E. Bruce Street
President
Texas Independent Producers & Royalty Owners' Association

I find it unusual to have to write this letter, but growing troubles in our industry make it necessary that I express my views and those of a great many true independents in the domestic industry.

Last week the Senate Finance Committee, under an amendment by Senator Williams of Delaware, finally brought forth a needed tax reform on foreign oil operations. As you know, international companies not only get full 27½ depletion on all their foreign operations, but in addition, foreign tax credits—which give them a big advantage over hardpressed domestic independents and other industries.

The amendment temporarily passed by the Senate Finance Committee would have stopped an oil firm from using any foreign tax credit resulting from percentage depletion to decrease its U.S. tax bill on non-mineral income in another country. When our self-styled industry leaders went into high gear, they persuaded Senator Russell Long of Louisiana (who really knows better) that the whole industry, including independents, oppose this foreign tax credit amendment. In our name Congressmen were told we oppose an amendment which did no more than cut out one minor way international companies gain tax advantage over domestic independents. Working behind Senator Long, who has the reputation as a friend of the domestic industry, they were able to convince three Senators to switch their votes, and soften the amendment to permit use of such tax credits, if they are used against income resulting from oil-related activity, such as refining or marketing. The result is a $10 million cut in tax credits instead of $40 million. No real victory was realized for the important principle of equal tax treatment between domestic and foreign oil investment by American companies.

February 6, 1964

Excerpt of Letter

TO: Mr. J. Ed Warren
Chairman
American Petroleum Institute

I realize that many of the majors feel as one major company official once stated, "There are more stockholders in my company than there are independents." This may be true, but the independents have been the backbone of the domestic industry for a great many years and they have not resigned their position yet.

April 7, 1964

Excerpt of Letter

TO: Mr. M.J. Rathbone
Chairman
Standard Oil Co. of New Jersey

Time is passing, and you have served your industry and your company well. Like all great oil men you unquestionably must step down from your role as the leader of your great company. Before you do, there should be time for you to reflect on the days you first started in this industry, worked your way up through the maze of competition to accomplish the leadership of the giant of all giants. Your good fortune with Standard of New Jersey is only a part of your success, for in your climb to prominence you have had the pleasure of rubbing elbows with thousands of independents who admire and respect you. There is no question that these thousands of friends in the industry ranks look up to you and your company's policies for the umbrella needed for the survival of a great independent industry.

As I close this letter I want you to know we are still hanging in there in spite of our derricks being battered, our tanks leaking, and not only is our equipment worn out, but our spirits and hopes as well.

June 10, 1964

Excerpt of Letter

TO: Johnny Mitchell

FROM: M.J. Rathbone
Standard Oil Co. of New Jersey

Your highly complimentary expression regarding the behavior of Jersey and its domestic affiliates is a source of great satisfaction to me. It is good to know that our managements have conducted themselves in such a fashion as to gain the goodwill and respect of men such as you, whom we regard as friends as well as competitors.

Over the years that you and I have known each other, there has been much change in the domestic producing situation. It is inevitable that change will always be with us. Keeping policies abreast of the times is a continuing and difficult responsibility of management. I am hopeful that our performance in the future will be regarded as in the best interests of the industry.

From time to time, I am sure, you are in New York. On one of your trips here I hope you might find it possible to drop in for a visit. Not only would it give us a chance to discuss some matters of mutual interest, but it would give me an opportunity to make you acquainted with some of my associates whom I would like you to know. My present plans are to be in New York most of the time during June and July, and I am hopeful that you might be here sometime during this period. If you would give me a little advance notice by telephone, I am sure we can arrange a mutually satisfactory time to get together.

June 10, 1964

Excerpt of Letter

TO: Mr. Fred L. Hartley
Union Oil Company of California

As a non-integrated independent oilman operating in California, I've reluctantly come to the conclusion that one big reason there is so little room in this state for active independents is that a handful of selfish major companies have a stranglehold on the oil industry out here. Making matters worse, it seems they're using certain so-called "independent" integrated companies like Union Oil of California to beat down any threat to their near-monopoly arrangement.

The bitter fruits of all this already include a depressed West Coast oil industry—one which could easily be flourishing in the otherwise booming economy of California. But what's more important, the long-range cost could include American lives—if we continue to let these selfish interests make California and the whole West Coast dependent upon highly-vulnerable foreign oil. Just because this arrangement is more profitable to importing companies, including Union of California, seems a poor excuse for the present sad state of affairs.

If the oil import program has any meaning at all, it is certainly not to empower a federal bureau to assure oil supplies from the most economical source and the most convenient and profitable one to the leading companies. It was rather to avoid American dependency on unreliable foreign oil.

It is a terrible situation when a handful of powerful major companies are allowed to run—and ruin—a defense-vital home oil producing industry. It is even worse when they can enlist the active help of a company which, calling itself an "independent," tries to infiltrate independent producer associations and keep independents from defending themselves. I say this is a disgraceful situation and that's why I've taken this means of calling your hand. Union Oil and the powerful interests you are fronting have as much to lose by such short-sighted policies as real independents. When the international majors have gobbled up most of us, Union Oil will be next—and federal control is certain to follow soon after.

July 23, 1964

Excerpt of Letter

TO: Honorable Ernest O. Thompson
Railroad Commission of Texas

It is my understandable pride and deep admiration for you that prompts me to write this letter. My purpose is primarily to commend you, in the warmest terms at my command, for your long years of unparalleled dedication and service to our industry, our state, and our nation.

When the history of this industry is finally written, you will tower above all others who may be listed among its architects and defenders. You are, in a very real sense, the father and the champion of the state conservation system which accounts for the greatness of this industry.

What makes your achievements all the more noteworthy is that, more often than not, you have had to fight our industry battles without the full support of all industry segments. Indeed, many of your greatest achievements on our behalf have been made over our opposition. The market demand proration system, for example, was viewed with grave suspicion by independents and some major companies alike. Yet it has provided the underpinnings of a stable and healthy industry, and directly made it possible to meet oil requirements of this nation in war and peacetime emergencies.

For you, we devoutly hope there are many fruitful years ahead during which your counsel and guidance of industry affairs will be welcome. Like so many other great industry statesmen, we will not always be able to expect you to continue in your accustomed active role of leadership. I speak, of course, of industry champions and friends of all segments, such as Walter Hallanan, Gene Holman, Jack Rathbone Follis, and others, must turn over the helm of your ship to some younger man. When this time does come, it is my hope that you will help our great Governor select the man to replace you.

As I end this letter, I would like you to know that my deep respect and admiration for you dates back to the first time I saw you soon after I started in this industry. As the years have passed, this respect for you has continually grown. I sincerely hope that the sands of time allow us to continue to enjoy our good friendship and to continue to work together for this great industry which is so indebted to you.

August 16, 1964

Excerpt of Letter

TO: Mr. John Swearinger
Chairman
Standard Oil of Indiana

I wish to thank you for taking your valuable time away from the American Petroleum Institute meeting and discussing the current energy problems we are now facing. The situation as we both see it is extremely grave for energy independence even though government and Congress don't agree.

The coming of the monopoly and concentration trend has brought almost a complete halt to the major companies' sense of responsibility to the industry. The ambitious selfish goals of these major international oil companies will not permit their industry responsibilities to interfere with the monopoly and concentration trend they are trying to accomplish.

The goal of the major companies seems to be to control as much of the overall industry as they can gobble up, disregarding completely how many small producers, suppliers, drilling contractors, refiners and small service stations are affected. All one has to do today is look at the shuttered doors of all segments of this industry that are going out of business each year. And yet we constantly hear from the major company officials that this is a united industry—that we should combine our thoughts and our efforts and work together, for there is room for both of us. Unfortunately, the independents now walking the streets find this hard to believe.

As far as majors are concerned, their prosperity will be bountiful. Imports will make stars out of all you chief executives. Special foreign tax credits will add to this prosperity.

John, our nation will never again be energy independent. You majors were given too much and too fast. There were great leaders among you majors. The likes of Jack Rathbone, Michael Haider of Jersey,

R.G. Follis, Carl Reistle, and I include yourself in this group. There was a great tendency among all of you to hopefully keep America strong in energy but the tide of profits from unreliable imports of foreign oil swept all of you off of your feet.

This was once such a great industry. Manned by independents and majors who stood ten feet tall in any crowd.

A few will still be around when the roof falls in. When that happens the same tall men won't be so tall anymore and there won't be any looking back to see how we got off track . . . it will be too late.

September 9, 1964

Excerpt of Letter

TO: Mr. Forrest N. Shumway
President
Signal Oil and Gas Company

Circumstances have caused me to write directly to you instead of to your department heads, who would normally handle the particular situation to which I am calling your attention. This matter was first referred to Emmet Walters and then to Russ Green, Jr., both of whom I know quite well. The problem at hand goes beyond their jurisdiction, for they are company officials concerned solely with company welfare.

In closing, I wish to convey this thought to you. The ownership of this lease in dollars and cents is insignificant to my company. My concern is rather the company loyalty that you expect of your employees and that we expect of ours, whether they be regular employees or those hired from the outside to assist us in our work. When your company interferes with the relationship between the small company and its brokers and feels free to deal with them while they are still under obligation to the small company, then we are traveling a one-way street—all yours.

In California particularly, brokers perform a major function for our company. The single-minded devotion of our brokers is as essential to the small independent as is the loyalty and confidence of your key employees to your organization. We expect your company and other major companies to respect our position in the same way that we respect yours.

September 22, 1964

Excerpt of Letter

TO: Johnny Mitchell

FROM: Forrest N. Shumway
President
Signal Oil & Gas Company

We have received a copy of a mimeographed letter dated September 9, 1964, addressed to Mr. Forrest M. Shumway, President, Signal Oil and Gas Company, and written over the facsimile signature of Jade Oil & Gas Co., Johnny Mitchell, President. The letter indicates that copies were sent to other persons and we have been informed that you have distributed additional copies to persons not named therein. The letter contains false statements, attacks our integrity, and in the opinion of our counsel is libelous and is also a slander of title.

Demand is hereby made that you mitigate the damages we have suffered through your libelous statements and your slander of our title to the lease by a retraction and apology and that this be given the same distribution as your letter to us.

October 9, 1964

Excerpt of Letter

TO: Mr. Forrest N. Shumway
President
Signal Oil and Gas Company

We are in receipt of your letter dated September 22, 1964. Although I realize that your primary concern was to justify the actions of your company, I am sure that you would not knowingly rely on misinformation to support your claims. Accordingly, I feel that perhaps you are not in full possession of the facts.

After June 1, 1964, I called Russ Green, Jr., Emmett Walters and Don Wagner, but their uncooperative attitude made it impossible to reach a solution. At that time I made it known to them that Jade was claiming the right to this lease because Mr. de Golyer was working for Jade. I must concede that quite often companies other than Signal have, perhaps inadvertently, infringed on the rights of others. However, the executives of responsible companies, when informed of the situation, are generally quick to rectify such errors.

We have reviewed this matter with counsel and have concluded that our position is just and correct, and that if Signal will not agree to transfer of this lease to Jade, we are prepared to file a lawsuit against your company in order to protect our rights. Your contention that you have been libeled is incorrect, and I might point out that I have been advised by counsel that truth is always a defense in a libel action. I have fought too long for principles to allow your company to abuse my company and the rights of independents. I know that unfortunately most independents, because of their size, have to accept, without recourse, the acts and dictates of major companies. On this occasion, we are prepared to litigate the issue and to defend our position before a jury in the proper court.

November 4, 1964

Excerpt of Letter

TO: Johnny Mitchell

FROM: Forrest N. Shumway
President
Signal Oil and Gas Company

Since my letter of September 22, and also as a result of your letter of October 9, 1964, to which this letter is a reply, a detailed investigation was made of the facts relative to Signal's acquisition of the 25 acre Dutch Slough lease from Transamerica Development Company through Mr. Arthur de Golyer, whom you have alledged was your agent in acquiring that lease for your company.

This investigation strengthens and affirms the position outlined in my letter of September 22, 1964, that the publication of your letters are libelous in nature and futher constitute a slander of our title to the 25 acre lease obtained from Transamerica Development Company.

I am, therefore, again demanding that you publicly retract the libelous statements published through your letters of September 9 and

October 9 with as much force and coverage as given the original statements.

November 13, 1964

Excerpt of Letter

TO: Mr. Ralph W. Trueblood, Jr.
Vice-President
Belridge Oil Company
President
Oil Producers Agency of California

In the first place, Canada itself is a big importer of oil. They import about as much cheap foreign oil on their east coast as they are forced to dump on us from their surplus producing western provinces. As a matter of fact, the same international companies who import here have it rigged to guarantee Canada that any time they increase their imports of Venezuelan or Middle East oil into Canada, they'll dump an equivalent amount of Canadian oil on the U.S. market.

It is true that Canada was given an exemption from import quota controls on grounds that this imported oil is safer because it moves by pipelines from Canada. But in an emergency, Canada would need all her own oil and more, especially if her imports were cut off or cut down.

June 9, 1965

Excerpt of Letter

TO: Honorable Nicholas deB. Katzenbach
Attorney General of the United States

In particular, I suggest that attention at this time upon negotiations throughout which Interior officials committed themselves to a special arrangement by which Phillips Petroleum Company may import some 50,000 barrels a day of products via Puerto Rico, without having to pay the normal excise tax upon imported products of this kind. This privilege involving a $600,000,000 petrochemical complex, was awarded this company by negotiation with Interior officials, without the benefit of competitive bidding. This means that product imports of other competitive companies will either have to be cut back to make room, or total non-residential product imports will increase. Also it will require Presidential sanction, since the Executive proclamation will have to be amended to allow for it. I feel sure the President will wish to consider this particular Interior deal more carefully because it does require his direct affirmation.

It is hard to believe Secretary Udall's reasoning that the private negotiations with Phillips in Puerto Rico was done in behalf of Puerto Rico's economy. Any reasonable man knows that Phillips, Standard, or any other major international company, is interested in the earning power for itself and its stockholders. We have witnessed the selfish ambitions of all these major companies and you can rest assured that Phillips' move is motivated to gain an advantage only for itself. I further believe that Secretary Udall does not have the basic background of oil and gas matters, nor the courage to grant such a permit which will unquestionably destroy the import program without advice from some other authority.

July 15, 1965

Excerpt of Letter

TO: Mr. Johnny Mitchell

FROM: Stanley Learned
President
Phillips Petroleum Company

Attached is a copy of letter written today to the Honorable Nicholas deB. Katzenbach, Attorney General of the United States, which is self-explanatory.

It is evident to me that your letter of June 9, 1965, to Mr. Katzenbach was written by you for the purpose of, and with the malicious intent of, doing damage to Phillips Petroleum Company. We deem your inaccurate statements and the press reporting which your letter inspired as harmful to Phillips. For this reason I wish to clear up any misunderstanding by sending a copy of my letter to each of the parties to whom you mailed your letter. Will you please, therefore, be kind enough to send me your list with addresses?

July 15, 1965

Excerpt of Letter

TO: The Honorable Nicholas deB. Katzenbach
Attorney General of the United States
Washington, D.C.

FROM: Stanley Learned
President
Phillips Petroleum Company

In a letter to you of June 9, 1965, Mr. Johnny Mitchell of Jade Oil and Gas Company made several statements and suggestions reflecting adversely upon officials of the Department of the Interior and Phillips Petroleum Company and requested that your Department undertake an inquiry into Interior's administration of oil-related matters. Only recently a copy of this letter was brought to my attention. While it is not necessary to respond to all of the inaccuracies in the letter, I feel it is my duty, on behalf of Phillips Petroleum Company, to cover certain points.

"It is hard to believe Secretary Udall's reasoning that the private negotiations with Phillips in Puerto Rico was done in behalf of Puerto Rico's economy."

Public testimony has revealed that the Puerto Rican petrochemical project will provide a very substantial boost for the Puerto Rican ecomony. We believe that those officials entrusted with economic development in Puerto Rico are better witnesses than Mr. Mitchell to the beneficial effects of the project. In this respect Mr. Rafael Durand, Administrator of the Economic Development Administration of the Commonwealth of Puerto Rico, stated the following at the hearing of July 31, 1964, with respect to the Phillips proposal.

Talk of a "$155 million government giveaway" in connection with the government's helium program has been strongly protested by Interior officials and the four contracting companies as untrue.

Moreover, the statement, erroneous as it was, referred to the entire purchases made under all four contracts and not just to those purchases from Phillips as implied by Mr. Mitchell

July 22, 1965

Excerpt of Letter

TO: Mr. Stanley Learned
President
Phillips Petroleum Company

I am enclosing a copy of a second letter I have written to the Honorable Nicholas deB. Katzenbach, restating my position in the matter of the Phillips-Puerto Rican decision.

It has not been my intention to single out Phillips Petroleum Company, or any other company, for condemnation, but rather to point out what I feel are the extremely dangerous consequences of the issue in which your company has become involved.

As an independent, I feel the responsibility to speak out against any decision that will have such a damaging effect on the independent domestic producing segment of our industry.

July 22, 1965

TO: The Honorable Nicholas deB. Katzenbach
Attorney General of the United States

FROM: Mr. Johnny Mitchell

Please permit me to comment on the letter to you of July 15, written by Mr. Stanley Learned, president of Phillips Petroleum Company, in which he challenges certain statements I made in my letter to you of June 9.

For the record, my letter of June 9, 1965, stands as is, and I here repeat my suggestion that you undertake an investigation of the special arrangement by which Phillips may import some 50,000 barrels daily of products via Puerto Rico. Such investigation seems all the more needed now to reveal whether my statement or Mr. Learned's more accurately reflect the situation. I welcome his taking up the challenge, and by implication, agreeing that the matter is of such gravity as to require your personal attention.

Allow me to deal briefly with those points in my letter which Mr. Learned chose to dispute:

As to his point number 1, Mr. Learned asserts there is no special arrangement by which these imported products will evade the normal excise duty, that the agreement is a matter of public record, and that Phillips was not accorded preferential treatment.

Gasoline imports into the United States are supposed to pay 52¢ per barrel duty. But by bringing this oil from Venezuela first into Puerto Rico, then converting it into gasoline, this duty is evaded. Naturally, no reference was made to this evasion in the published documents, but the fact is that these imports do evade the existing excise tax which Congress intended such product imports to pay. Had this oil been produced in Puerto Rico, escaping the duty would have been one thing; but employing this circuitous manner to evade it is another matter entirely.

As to point number 2, Mr. Learned takes offense at my question as to whether this award actually resulted from concern with Puerto Rico's economy.

The interest of Puerto Rican officials in this deal may well be genuine, and it might be a fine thing for the Puerto Rican economy. That's beside the point. My point is that arrangements of this kind to undermine the imports program are a dangerous and improper means of helping the Commonwealth of Puerto Rico. I certainly have no objections if Phillips undertakes to build any kind of facility they desire anywhere in the world, and I hope they build two petrochemical plants in Puerto Rico. But I don't want to have it partially subsidized by special privileges to import products here, primarily at the expense of the domestic independent producers. If this sort of thing can be excused on grounds that it helps Puerto Rico, then why not allow a similar arrangement for a petrochemical plant in Appalachia, or Oklahoma or Texas.

My point was simply that the real motivation for this arrangement is to assure an economic windfall for Phillips Petroleum Company, not Puerto Rico, and I do not believe that should occur at the expense of a defense-vital home industry.

As to point number 3, Mr. Learned cites Interior Department assurances that the move will not be allowed to cripple the imports program.

I say to you that any such assurances on the part of the Secretary and his assistant serve only to show how unrealistic and naive they are in approaching this whole matter. Unless Phillips alone is to be allowed this special privilege, there simply cannot be a corresponding cutback of other product imports to keep total products from rising. There just aren't enough product imports that will result from this loophole. It's plain silly for Interior people to keep telling Congressmen and the public that they can cut new loopholes in the program and then offset them some way or other without crippling the program.

September 9, 1965

Excerpt of Letter

TO: The Honorable Russell B. Long
U.S. Senator

I appreciated hearing from you very much and I want to thank you for your inquiry into the Phillips-Puerto Rican issue. I, too, regret that your effort was fruitless. A number of inquiries into Interior's actions were made by other Congressmen and Senators resulting from my letters to them requesting such inquiries. In each case, the feeble answers they received from the Cabinet offices were shameful.

I predict that time will prove that the Phillips-Puerto Rican decision was a top priority matter, forced on the Interior Department. We have the suspicion that this package was approved before the knob of Secretary Udall's door was ever turned. Mr. K.S. Adams is recognized as a very close friend of Abe Fortas and Oscar Chapman, and what obligation the Administration owes Mr. K.S. "Boots" Adams has never been disclosed. At any rate, the result is a hell of a giveaway.

Many people seem to assume that our President, whose roots are deeply embedded in Texas, is not going to let anything happen to injure Texas and its industries. Even though he was reared a Texan, he is now our President, with many other obligations to serve all of the states for the welfare of our nation and its millions of citizens. This obligation to serve the nation as a whole necessarily involves matters that seriously affect Texas and its industries. A decision that affects Texas should be the business of our Congressmen and Senators from Texas. The weakness in the Texas Delegation is that most of our Congressmen and Senators are not inclined to differ from the President, a Texan, on his proposals and forget that they are supposed to defend the best interests of the State of Texas.

September 13, 1965

TO: Mr. Johnny Mitchell

FROM: Mr. Jack Valenti
Special Assistant to the President

You and I have been friends a long, long time. My affection and respect for you remains undiminished. That's why I was surprised at the rather bizarre statement you made "what obligation the Administration owes K.S. Boots Adams has never been disclosed." There is implicit in that statement a kind of mysterious alignment between Mr. Adams and the government, in which Mr. Adams seems to be a favored person. This is ridiculous as well as totally untrue. Of course, you know this.

September 23, 1965

TO: Mr. Jack Valenti
Special Assistant to the President

FROM: Mr. Johnny Mitchell

You and I know that I'm no diplomat, and I don't try to be. I call 'em just as I see 'em. I strongly feel that's what the President needs most from his friends. The worst disservice we could do him would be to say that whatever his Cabinet or his Administration does is right, or shouldn't be criticized because he is a Texan or because he understands our industry or because he's a friend.

The President, even as effective as he is, and make mistakes, and he will get blamed when some in his Administration make mistakes. I think a mistake is being made in the Phillips deal. It's bad, it's unfair.

September 30, 1965

TO: Mr. Johnny Mitchell

FROM: Olin E. Teague
Congressman

Yesterday, listening to a number of Members from Texas discussing the depletion allowance, your name came up; and the remark was made that someone, somewhere may be able to tell Johnny Mitchell something, but they are not going to be able to tell him much because he already knows it all. The people talking did not know that I was acquainted with you. The people talking were people who are very interested in maintaining the depletion allowance. The people talking were, on the whole, people who have opposed the Administration on the majority of the issues in the Congress so they were not among the group whom you say is "the weakness in the Texas Delegation" inclined to differ with the President, a Texan."

October 15, 1965

Excerpt of Letter

TO: Mr. Jake Hamon
Dallas, Texas

I have just returned to Houston and become aware of your letter of resignation to TIPRO. I am not as concerned about your resignation as I am about your feelings on independents expressing their opinions, either emotionally or with great deliberation, on foreign tax credits, foreign depletion, or any other industry problem that effects the independent.

I can assure you that I have as great a love for this industry and hopes for the industry unity as anyone else. You have the same feeling for you have mediated the cause of the independents and have been the buffer between the independents and majors for a great many years. However, your independent ties are coupled with a responsibility to the majors. I do not feel as strongly as you in this respect, nor do I feel the same ties and responsibility to the leaders of our government as you. Because of your feelings toward each of these factions, they constantly call on your judgment and abilities to smooth the troubled waters between independents and majors, and it is only natural that you become deeply concerned when issues arise that are certain to rock the boat of unity.

I, too, feel a definite responsibility to the industry. However, my purpose of my obligation is dedicated to the survival of the independent producers.

Our good friend, Michel Halbouty, took up your cry. He, too, is a good friend of mine; yet Michel, like yourself, has included an obligation to major company officials as a part of his requisite for success. It takes all breeds of independents, with their various philosophies, to succeed. Each has his right to fight for the principles he believes in.

December 31, 1965

Excerpt of Letter

TO: Mr. John Doe
Washington, D.C.

I believe the time has come for somebody who isn't running for or from anything to talk frankly about the oil industry mess in Washington.

If you're wondering why I'm addressing this letter to you, Mr. Doe, it's because everybody in a position of authority there is either swearing off oil issues or has demonstrated an incapacity to understand what's happening. Let me tell you what's happening to the independent oilman.

I think the oil industry has been the whipping boy of the politicians long enough, and that the President owes it to the industry to face them down once and for all. What could the President lose? Plenty, if he's making oil policy on the basis of what's good for America.

Well, anyway, that's why I'm not writing the President these days. You may ask why I don't write some oil-state Congressmen. I tried that too. Found they're all afraid somebody might get the idea they're criticizing the President if they do anything that hasn't been cleared.

A few months ago, in frustration, I decided to write the Attorney General, on the naive assumption that he'd be interested in hearing about the monopoly trend in this industry. I thought he'd also like facts about a little situation that's going on that could do for some selfish oilmen what Teapot Dome did for old Harry Sinclair . . . get them jailed.

After the Phillips fiasco, the petrochemical people got wind of a possible quota handout, and began threatening to build plants abroad if they didn't also get a quota subsidy. Clinched their case when they added that handing them this subsidy would increase their exports, and help the balance of payments. Mr. Udall apparently thought this was Great for Society. So he broadened the allocation base again to include petrochemical plants. Only Phillips is happy, though hopeful there will be no Senate investigation. The rest aren't happy. They claim his handout wasn't near big enough to keep them from building their plants abroad and won't subsidize more export sales after all. So all we got was an increase in imports to appease people who don't aim to be that easily appeased. Meanwhile, Mr. Udall got some temporary mileage out of some loose talk at his press conference about "freezing" something or other, and a few IOCC people who hadn't done their home work thought he had frozen the level of imports so domestic production could get a little growth.

About this time I got a notion that independent producers might get included in the Great Society. Since Mr. Udall is so fond of imports, and quotas seem to be his answer to everybody's troubles, why not include the people Congress intended to protect by import curbs in the first place? Sort of "go along to get along," if I may coin a phrase. Well, I developed a plan to do this and sent it to all the major company brass. Got a lot of interesting answers, all adding up to the same thing: "Johnny, if you don't stop messing around, you're going to screw up our whole imports program." Here I'd been operating on the theory that they'd beaten me to that.

So what about it, you may be asking. What if the independent producers are going under during peak prosperity for the nation? After all, they're just oilmen, and all oilmen are rich anyway. I'll tell you what about it, Mr. Doe. When the independent oilman is hurt, the domestic wells don't get drilled and the oil doesn't get found. Let this go on a few years, like it has, and our reserve productive capacity goes to pot. If any significant portion of foreign oil is cut off to our allies, like in the Suez crisis only a few years ago, we'll be helpless. It's that U.S. reserve production capacity on which the free world "floated to victory" in recent wars. And it's going fast. So just cause the majors are making lots of money doesn't mean the oil industry is doing all right, Mr. Doe. Wouldn't take much to correct the situation, but sure doesn't look like Mr. Udall is going to bother about it.

We have other troubles, Mr. Doe, and I may tell you about them in my next letter. Meantime, if you run into the President, you might just casually mention that we aren't very happy.

May 29, 1966

TO: Mr. Carl Reistle, Chairman
Humble Oil and Refining Company

Carl, back in the 60's there was a lot of territory separating a small independent and the Chairman of Humble. Why you took an interest in me, I still don't understand. I was always a problem to all majors with my speeches, usually against government and major oil companies. I won't forget my caustic speeches at Commission hearings, and my usual remarks about major oil companies at the National Petroleum Council meeting in Washington. I was always concerned about your reaction to my outbursts and believed that you would eventually give up on me. You never did Carl, at the end of each meeting in Austin and in Washington you would have a folded note handed to me quietly in the crowd. The note would usually say . . . "You are the dumbest S.O.B. I've ever known. You'll never learn to keep your mouth shut. My plane leaves at 3:30 p.m. Be there." Then on board, in front of all, you would preach to me like a father to a spoiled son.

You always ended up saying . . . "Who in the hell do you think you are: . . . You can't control Congress . . . You can't change the thinking of the Executive Branch . . . And you can't influence the majors . . . Why don't you give up?" I never believed you really wanted me to stop, or I would have lost you long ago.

July 12, 1966

Excerpt of Letter

TO: Honorable Thomas H. Kuchel
Washington, D.C.

I read with interest your demands for an explanation from Secretary of the Interior, Stewart Udall, for the movement of 10,000 barrels of gasoline, duty-free, from Puerto Rico to the West Coast. Those of us in the oil industry, who have long waged a lonely and losing fight against detrimental governmental policies, welcome your aid. You are right when you say this latest importation of 10,000 barrels daily of gasoline jeopardized California's competitive marketing economy.

Assistant Interior Secretary J. Cordell Moore's glib denial of your charges is just another indication that this Administration is determined to keep, regardless of cost to the domestic industry, a deal it made to give extreme preferential treatment to companies . . . one in particular . . . which import their foreign oil here through Puerto Rico. Mr. Moore will give you no satisfaction, because he cannot. If you simply want to make an impression on California refiners, you can carry on your correspondence with Interior officials who really don't have anything to say about policy at that level. But if you are serious in wanting to do something about a situation that is already bad and due to get much, much worse, then I will tell you how.

In case you are not aware of it, the future problem that will make your present 10,000 barrel import into California look like a drop in the bucket relates to a deal made with another Republican. When the results of that arrangement are realized soon, imports of gasoline from Puerto Rico will be many times the present level . . . . enough, in fact, to hurt the whole price structure of the domestic oil industry. I am talking about Boots Adams, chairman of the board of Phillips Petroleum Company, who, with some well-placed insiders, walked off with the biggest prize ever handed any company or individual, including the one that led to the Teapot Dome scandal.

A number of us asked serious and probing questions before this deal was signed by the President. Almost all the other major integrated companies declared they would want a similar plum, if Puerto Rico was going to be made the playhouse for Phillips. As you know, Puerto Rico doesn't produce any oil. It's just a convenient port of entry for foreign oil, and when refined there, escapes the 55 cent per barrel gasoline tariff that would apply if imported direct from other countries.

You have demanded an explanation from Secretary Udall, and got doubletalk from an assistant secretary. You can tell the California refiners you tried, and quit . . . . or you can do something which can get the job done, and thereby help the whole domestic industry.

October 12, 1966

Excerpt of Letter

TO: The Honorable Craig Hosmer
Washington, D. C.

I have reviewed the speech you made recently in Long Beach, California with great interest, and commend you on your knowledge of the problems of the oil and gas industry in California. Your criticism of me has merit, for I, like many others in the oil industry, expect our Congressmen to know all the answers without presenting the facts. We know our problems, but it is our short-coming in not taking the proper steps to explain them to our Congressmen. We appreciate your taking time to find out the facts for yourself.

You may not be aware of it, but since your speech you have joined the ranks of other distinguished industry statesmen. To qualify for this privilege, you have to be criticized by California's most active industry association, Oil Producers Agency of California. I read this agency's comments on your speech in The Oil Daily the other day, and you are now elevated to the same statesmanship status in our industry as Jack Rathbone, past Chairman of the Board of Standard of New Jersey. He, among others, has also been criticized by this same agency when anything is said to disturb the status quo situation of the California producing and marketing industry.

In his published memorandum dated September 14, 1964, Fox claimed that since most of the increase in imports comes from Canada, there's nothing to worry about. He says the Federal government considers Canadian oil as domestic oil insofar as national security and imports policy is concerned.

I say that no knowledgeable government official believes that. You don't believe it, I surely don't believe it.

In the first place, Canada itself is a big importer of oil. They import about as much cheap foreign oil on their east coast as they are forced to dump on us from their surplus-producing western provinces. As a matter of fact, the same international companies who import here have it rigged to guarantee Canada that anytime they increase their imports of Venezuelan or Middle Eastern oil into Canada, they'll dump an equivalent amount of Canadian oil on the U.S. market . . . most of it into District V.

It is true that Canada was given an exemption from import quota controls on grounds that this imported oil is safer because it moves by pipelines from Canada. But in an emergency, Canada would need all her own oil and more, especially if her imports were cut off or cut down.

February 13, 1967

Excerpt of Letter

TO: Mr. R.C. McCurdy
President
Shell Oil Company

It isn't very often that a small matter such as this be referred to you. However, what is routine in the daily workload of an international company is important enough to me as a small independent to write to you.

Just last week the National Petroleum Council came out with the finding that the independent producer, once the historic oil finder and wildcatter, is no longer a key figure in the domestic oil industry. There are reasons for the independents' problems in this industry. The Administration and the majors share the blame for premeditated handouts, unfair marketing practices and excessive imports. These are the factors that contribute so heavily to the low price of a barrel of crude.

What also hurt the independent is the uncooperative attitude of major companies. This has done as much to reduce wildcatting and exploration development as the price of crude. Major companies at one time were receptive to the independents' efforts to explore, were always helpful in either dry hole contributions or acreage contributions. They also considered that the cost of the venture should be proportionately shared. These policies have changed, for now major companies not only decline to cooperate, but hardly ever share in ventures of mutual interest.

I am writing this letter to you not so much to make an issue with your company but, as an example, explain to the industry that one of the primary reasons for the independents' decline is that the major companies are not sharing their part of the risk to explore, signifying they are not concerned enough with the nations declining reserves and the nation's security.

Where should the responsibility be placed? Only one place . . . . . . on each executive officer of each major company.

February 27, 1967

Excerpt of Letter

TO: California Delegation
United States Congress
Washington, D.C.

FROM: California Independent Producers
& Royalty Association
(CIPRO)

It is with deep concern that we question the right of the Administration to threaten with reprisals the industry that has devoted 100 years to our consumers' needs and has been ready in every national emergency, fully prepared to serve our nation. Does our Secretary of Interior have the authority to threaten us with open-door imports of cheap, unrealiable foreign crude, along with tax-exempt imports of gasoline? Does the Under-Secretary, Charles Luce, completely without knowledge in petroleum affairs, have the right to further blackmail a free enterprise system?

There may be other reasons besides threats and reprisals that will cause some of the majors not to join in the price hike. There is always the chance that some ambitious chief executive, regardless of his personal beliefs and standards, might accept a "special handout" for his company as compensation for "going to bed" with the bureaucrats.

May 15, 1967

TO: Mr. John Doe

I have to blame the President because of two things. First, when we had a chance of getting a good import law in 1958, when he was Senate Majority Leader, when we had a chance of getting the 16.6% limit of domestic production on imports and products placed into law so the bureaucrats couldn't twist it; Senator Johnson would back us only enough to get some mealy mouth words in the law which the bureaucrats would distort at will.

Why would the President permit Secretary Udall to hand out import quotas in Puerto Rico where his close friends, Able Fortas and Oscar Chapman were both involved. Why is he about to do the same with friends like Hess and Tommy the Cork in the Virgin Islands. The silly argument that these handouts are to resurge their economy is the flimsiest excuse that has ever been offered to the American public. Even the natives of those countries don't take that seriously.

October 26, 1967

Excerpt of Letter

TO: Honorable Stewart Udall
Secretary of the Interior
Washington, D.C.

The decision of whether independent oilmen remain as an integral part of the domestic industry depends upon immediate action on part of government acted on by your department. You must be realistic about the many problems which began a few years ago but have finally come to a head under your authority.

Your decision to open floodgates of foreign oil in such proportions as to threaten the future of the oil domestic producing industry has to be judged against you and our President Johnson. At the present rate of special handouts and now the establishment by you of special trade zones, imports are destined to reach a level of about 30%. This continued increase in imports from the Middle East to special interest groups will some day make the Arabs block the masters of this troubled world.

We are subjecting ourselves to international blackmail possibilities with our own current policies in the Middle East. It is quite conceivable that some day the Arab nations will force the United States to choose between the Middle East and Israel. With so much money and profit involved to a selected few international companies there is no doubt in my mind whose side the United States will choose.

We are in essence paying through the nose for foreign oil and waiting for the day when we must buckle down to Arab demands.

July 8, 1968

Excerpt of Letter

TO: Honorable Ronald Reagan
Governor of California

We have never met, and by way of introduction, I served as the first president of the California Independent Producers & Royalty Owners Association in 1966, as President of the Texas Independent Producers & Royalty Owners Association in 1963-1964, and served as a member of the National Petroleum Council during President Kennedy's administration.

As a result, I feel I know the working attitude of the major integrated oil companies, both in California and in other producing areas of the country. What I am concerned about is the monopoly trend of these major integrated companies in their operations in the State of California. This to me is all more strange since the same majors don't exhibit this same trend in their operations in other areas of the United States.

I am, therefore, taking the liberty of asking that you take time from your important work to examine a problem which, no doubt, you would much prefer left to others. But the matter of artificially depressed crude oil prices in California does concern every citizen of our state, and I believe sincerely it is one which can only be dealt with effectively if the Governor of California personally demonstrates that he does not approve of the monopoly-like conspiracy which is depressing crude prices here.

Should you advise me that you will also forward my letter to the State Lands Commission for their investigation of this matter, I must tell you, Governor, that the State Lands Commission in the past have been too involved with major oil company executives to render impartial judgment. These major oil companies are quite wise in dividing the State Lands Commissioners among themselves so that each commissioner has only a few oil companies to favor. Since the monopoly conspiracy of these companies in California is single purpose—"low crude prices"—the State Lands Commission is not apt to apply any pressure on their good hosts, the major oil companies, on this most important issue.

So long as major company executives are good political contributors, many in political life are unwilling to challenge their business practices, whatever the cost to principle and whatever sacrifice to State revenues. Such actions as this may not be the most popular thing you could do politically, but in my judgment it is the right thing to do and should be done.

November 18, 1968

Excerpt of Letter

TO: Honorable Ronald Reagan
Governor of California

In answer to my letter to you on July 8, 1968, I received a letter dated August 29, 1968, from your Director of Finance, Mr. C.W. Weinberger. Instead of answering him at this time, I prefer writing you again with a copy to Mr. Weinberger, which will save you the trouble of forwarding this letter to him.

The State Land Commission recently released information regarding the long-awaited, expensive, State-Paid DeGolyer and MacNaughton "Report on Crude Oil Pricing in California." The State Land Commission could have gone directly to Standard Oil of California Chairman Otto Miller for the same information, free of cost, for this is undoubtedly where DeGolyer and MacNaughton had to get most of the information for their report. This type of work is not DeGolyer and MacNaughton's cup of tea.

It is public knowledge that you are being advised not to get involved in any oil issues in the State of California. This is probably one of the reasons Mr. Weinberger asked me in his letter to go to the Justice Department in Washington if I had any question on monopoly in crude pricing as apparent only in California. More so, when only the State of California is being deprived of hundreds of millions of dollars. Of more concern to Mr. Weinberger should be the fact that these millions of dollars of the taxpayers' money winds up in the treasury of a few California-based major oil companies.

It is also public knowledge that you do have oil advisors. Unfortunately, this advice is major-oil-company-oriented. You also have one key advisor, a man of prominence in Los Angeles, who is close to the major oil companies and has been a servant of their needs for a great many years. It is hardly likely that his advice will be in opposition to the major oil companies and their unfair practices in California. His ties are too deep, his obligations too strong, and his knowledge of the independent oil man's problems too limited.

Each President and Chairman of the Board of these major oil companies understand one code . . . . to serve only their company's welfare and that of their stockholders. Your personal advisors on oil are out to serve themselves and the major oil companies who have been their master for too many years.

November 20, 1968

Excerpt of Letter

TO: Mr. Houston Flournoy
Chairman
State Land Commission
State of California

I have recently reviewed the the full text of the DeGolyer and MacNaughton Report on California crude prices which was authorized by the State Land Commission at considerable cost to the State of California.

I must compliment the State Land Commission for being consistent. Consistent, that is, in wasting the taxpayers' money.

November 7, 1969

TO: Mr. Michael Haider, Chairman
Standard Oil Company of New Jersey

The pleasure in being able to know you and Mr. Rathbone has been a special high point of my life. I have locked up in my memories three great industry leaders, all belonging to the same Standard of New Jersey family. They are Carl Reistle, Jack Rathbone, and Michael Haider. In the history of Standard to come, this combination will never be duplicated.

December 2, 1969

Excerpt of Letter

TO: The Honorable Wilbur Mills
U.S. Congressman

May I make one last appeal to you as a key member in formulating the final draft of the present tax reform bill.

Having had the privilege of discussing with you the merits of various tax proposals levied against the oil and gas industry, I believe some could be modified to fit into the tax reform without damaging this vital industry which is so essential to our national security.

We both recognize that as far as the independent oilman and his small investor are concerned, the most essential tax incentive that will keep us able to continue to explore for needed reserves is the intangible expensing feature of our tax laws. Of lesser importance is depletion.

May 7, 1970

TO: Mr. Clinton Engstrand, Chairman
Liaison Committee of Cooperating Oil & Gas Association

Secretary of Interior Hickel proposed an import program. His won't work. Not only has he added new problems, but he is recommending all the bad points of the present program.

He failed to eliminate the 200,000 plus barrels of handout, exempted imports going into Puerto Rico and the Virgin Islands. To make matters worse, he continues to allow 75,000 barrels per day of duty free gasoline, refined from this crude, to enter all market areas on the East Coast, West Coast and Gulf Coast. He should disclose that cutthroat marketers can purchase gasoline from Puerto Rico for any port in the U.S. at a price per gallon always a few cents under the domestic market price. This disadvantage not only riddles fair marketing practices, but the present program as well.

September 7, 1970

Excerpt of Letter

TO: Honorable Emilio Q. Daddario
U.S. Congressman

Since a remark by you has drawn a sharp response by a former assistant Interior Secretary, J. Cordell Moore, in the August 3, issue of The Oil & Gas Journal, I believe it wouldn't be out of order for me to direct a few words your way.

As you will recall, the point I was attempting to make to my Texas friends was that it is high time the import program was changed. Now is the time to allow the independent producers and explorers some kind of direct participation in the benefits of the quota allocation system.

This nation is currently producing all the oil it can produce on a sustainable basis, and is living off the shelf so far as the future is concerned unless there is restoration of former drilling levels. Today we are drilling about one half as many wells as we were when the import program was put into effect, and continuing to lose ground. Why? Because the import program doesn't do what it is supposed to do—and neither the Administration nor Congress seems to want to face up to that.

It isn't hard to explain, as the result of economic manipulation by major companies, this great nation could degenerate from the world's leading energy producing nation to one dependent upon unreliable foreign sources for our daily energy requirements.

July 24, 1971

Excerpt of Letter

TO: Mr. H.J. Haynes
President
Standard Oil Company of California

Over these many years, I've watched industry leaders climb the ladder to the top as if they were handpicked by Fate. As each reached the top, there was always our hope that at least one would detour from the competitive scramble for profit just long enough to see how badly the industry is faring, especially for the independents in California.

The room at the top has always been crowded with outstanding executives, usually breaking bread over their company's successes.

Each will tell you in private the same words, "I remember the early days of this industry, the helping hand of the major oil companies toward an independent." "I would really like to take time to review the many problems and help the small independents." "If I do, the other international giants will race right by me." "I can't ask for a reduction of imports for my company, for if I do another will get it."

Still Bill, with this competitive attitude now shared by all, the end result will be that doomsday will be here before you retire. And like the rest, you will be looking back and wondering why your company didn't help the small independent in their struggle for survival, and then you will fold your tent and like the leaders before you, silently slip off into the darkness of the night.

Whether it is too late for the independent segment of this industry to regain its role, I don't know. Incentives depend on future understanding between the independent segment and the major international oil companies, even more than on better "public" or "governmental" relations.

At best, it's a race for our lives to see whether this great nation goes under for lack of energy.

At the present rate, we are heading for a crisis that will shake the world. When that happens, there no longer will be any priorities from government for Standard of California, nor for Jersey Standard or the rest of the majors. You will be joining us independents as we slide down the tube together. And most tragic of all, our nation's dependency on foreign oil will become a true way of life here in America.

February 10, 1972

Excerpt of Letter

TO: Mr. Peter Flanagan
Executive Office of the President
Washington, D.C.

The oil industry is nothing more than the sum of the individuals who make it their way of life.

The nation depended on these individuals. Once in abundance but through people misunderstanding and government controls these ranks have thinned to only a handful.

Under present policies the petroleum industry will not turn upward, and certainly won't sweep anybody anywhere.

A few individuals will succeed, others will fail, and the sum of their limited successes minus their failures, will be summarized as the nations progress in our current energy crisis.

We're not going anywhere in finding new oil and gas reserves and our nations energy needs are increasing. That's why I am writing you.

June 10, 1972

TO: Honorable George Wilson
Executive Office of the President

Only recently have our industry statesmen permitted the true story of reserve productive capacity to be told in official documents. Long after they knew better, they were publishing figures showing the U.S. had well in excess of a million barrels daily unused capacity. The only thing resembling an excuse I've found for this tomfoolery is that they thought the OPEC nations were stupid enough to believe the U.S. could retaliate should Arab countries shut off our supplies or hike the prices too sharply. But we were only fooling ourselves. I suspect the Arab potentates knew more about the factual situation than certain leading brass of American companies.

April 30, 1973

Excerpt of Letter

TO: Mr. D.F. McKeithan, Jr.
Chairman
Liaison Committee

The real guilt of these same majors is that while they lined their purses with profits of the import program, they turned their backs on domestic exploration and development. In 12 short years they selfishly watched this great nation slip from a leading power to become a second-rate energy beggar.

The downfall of the independent producing industry will never be cataloged. Most of the blame is due to our own apathy and false "pride" of independence.

It takes gutty leadership—we can't win by keeping on extending olive branches and pulling back bloody knuckles.

October 31, 1973

Excerpt of Letter

TO: Honorable Henry Jackson
U.S. Senator

I am taking the liberty of writing you about a critical situation that is already present and will only worsen as time goes.

For the first time in twenty years the price of crude oil and gas is expected to reach a price that will rejuvenate domestic exploration and development of oil and new reserves.

Now that the independents can forecast a realistic price, we have increased both our exploration and development activities. But now, unexpectedly, we face a new serious problem; a serious shortage of tubular goods. This alone can put the brakes on oil and gas exploration programs vital to our nation.

Independents don't stockpile tubular goods for the simple reason that they do not have the capital. Independents, who actually drill three-fourths of all domestic wells, acquire their tubular goods from supply houses as they drill, at the time they are actually drilling, and require casing and tubing to complete their wells.

I am not pointing the finger at any company for stockpiling, or charging that any major mill has favored these companies. In normal times, the right for mills to sell to whomever they please should be their privilege. Yet in the face of our severe energy crisis, when our nation requires every barrel of oil and every cubic foot of gas we can produce, I feel that any company in the national interest should be willing to put part of its excess back into stock to allow the drilling of more wells now. I also believe that major mills would be agreeable to some form of allocation and distribution procedures, provided that the urgency of the situation is called to their attention by someone who is concerned with our nation's welfare.

January 2, 1974

Excerpt of Letter

TO: Mr. Robert O. Anderson
Chairman
Atlantic Richfield Company

When you call for elimination of percentage depletion allowance, you are also speaking for the independent producer who has accounted for some three-fourths of the oil and gas found in this nation. So let the record show that you do not speak for the typical domestic independent oil and gas explorer.

It very well may be that you and Arco President Thornton Bradshaw have figured out some way Arco could gain competitive advantage over other major petroleum companies through this maneuver. Maybe Arco is embarrassed by its earnings picture, doesn't need improvement in its earnings in order to carry its share of the load in making available more energy resources for our nation. But domestic independent producers, unlike the international giants, have not been enjoying record earning and growth in recent years.

I realize you can't be serious, of course. But the damage to our industry could be immense from this kind of childish behavior. It is one thing for this suggestion to come from someone in a position to assure us a free market price for our product in exchange for our yielding up percentage depletion.

It is quite another for executives of a leading international company to do so. All you can possibly hope to do is confirm the erroneous impression of our industry's professional critics that depletion is unnecessary and unjustified.

If you are intent on volunteering something to appease the critics, let me suggest that you show real statesmanship and volunteer your foreign depletion and tax credits. This would be entirely logical at this time. There no longer exists any justification for permitting American companies to classify royalty payments to Arab nations as income tax and then deduct this on a dollar-for-dollar basis from taxes owed the U.S. Government. The need is not for more Arab oil development, but rather to bring American international companies home to help independents look for oil and gas within the U.S. We need resource development at home, not abroad. Eliminating foreign tax credits and foreign depletion at this moment in history would be a wholly constructive source.

April 23, 1974

TO: Honorable Henry M. Jackson
Senate Office Building

Chairman Wilbur Mills and his wrecking crew heard a lot of testimony from the industry and the Administration, all pointing up the fact that tax laws had been written to encourage too much oil development abroad and too little at home. So what happened? They came up with a bill which would impose additional taxes during the next six years on domestic production of $13 billion and on foreign production of only $3 billion. What's more, the "windfall profits" tax they propose is a flat excise tax—not on the international companies which have gotten the windfall profits but upon domestic crude. All this adds up to an attempt to get—in a far more damaging way—the crude price rollback that you were seeking in a direct manner. I don't think either course is good legislative policy, but there's absolutely no question but what your proposal would be better for the nation than the silly thing that Mills' Marauders have come up with.

On depletion, the executives of some of the major companies concluded that they were so ashamed of their "obscene" profits that they advocated getting rid of this "albatross around our necks." They weren't speaking for us little independents. We need depletion and intangibles to keep finding oil. Nevertheless, what did the Ways and Means Committee do? It went along with the majors—providing a phase-out of depletion for all of us. The only concession made to us independents was that we could keep a 15% depletion rate a few months longer than the majors.

This is no time to shoot from the hip, something I believe the Congress has been doing too often lately.

If we don't find answers to the energy problem, there won't be enough left of this nation to justify all the efforts you are making to the Presidency.

June 27, 1975

TO: The Honorable Bob Eckhardt,
U.S. Congressman

Bob, it takes a pretty good American to admit a mistake in judgment. One that shuts the door to the future energy security of our nation.

I've walked down the halls of Congress these past few years, and I've met a few such great Americans in the Congress. Your sensible colleagues will respect you for facing hard economic facts, those that could encourage the independents to step up their exploration and development of new domestic oil and gas.

December 5, 1975

Excerpt of Letter

TO: The Honorable Frank Zarb
Administrator
Federal Energy Administration
Washington, D.C.

As an independent oil producer who has held you in high esteem I must express deep disappointment in the course you have taken with regard to the horrible omnibus energy measure.

Upon realizing you had been trapped into agreeing with the maneuver by Senator Jackson and his cohorts, instead of acknowledging you made a serious blunder and so advising the President, you've sought to protect your own flank. By attempting to justify your course in political terms, implying that this is necessary to the re-election of President Ford, you've asked the President to compound the error by ignoring his other advisors and joining you in approving the bill.

If you prevail, and if the President signs this terrible bill, you and he must share a great responsibility for undoing all the good you've done in the past. Up to this point you've stood firmly against the demagogues led by Senator Jackson. Now suddenly you're letting them call the tune, and implying that you have no alternative. There is a reasonable way out of this, consisting in the extension of the present arrangements for the time being—and I don't believe doing what's right will prove to be bad politics in the end. On this one your dead wrong, Frank, and if you are the man I think you are, you'll stop compounding your error. By urging the President to join you in your mistakes is a course unworthy of a fine public servant such as yourself.

I think I would be remiss in my duty if I did not tell you I sincerely believe that should this bill become law, it'll take a very short time to realize how disastrous it is. Unless the government wants to put oilmen in uniform, I don't believe they will continue to produce their oil at less than replacement costs. We did exactly that for a good many years hoping for an awakening as to what was happening. Now that the realization has come and President Ford has undertaken to do something about it, we do not propose to let it happen again. Not only will we not carry out our drilling program, but even current production will be cut sharply if the President goes along with the outrageous maneuver by Senator Jackson and his cohorts.

March 9, 1976

TO: The Honorable Bob Eckhardt,
U.S. Congressman

Back when the Texas State Legislature attempted to assert limited jurisdiction over natural gas prices—in order to end the rankest kind of discrimination against small producers, royalty owners, and the State of Texas—the pipeline purchasing companies fought us bitterly. Their battle cry then was "sanctity of contracts," you will remember, saying that where they had contracted for gas for as little as 2¢ per thousand cubic feet no governmental body should be allowed to interfere with the "Sanctity" of those contracts. They won that battle in Texas—with the result that many got trapped under federal regulation at absurdly-low prices.

Now the question before Congress is whether gas under contracts which expire should by law be locked into those purchasers forever at regulated prices. No longer do we hear the gas pipeline companies defending the principle of "sanctity of contracts." Instead, they maintain that to protect the public they must be allowed to continue taking that gas at the old regulated price, even though the contracts under which they originally bought the gas no longer exist. This is not surprising. The surprise to me was to hear you take up their battle cry on the floor of the House of Representatives during the recent debate on gas deregulation.

Letters

July 28, 1958

TO: Mr. Johnny Mitchell
Houston Club Building
Houston, Texas

Dear Johnny:

When I was in Washington recently, I was distressed to learn of your public criticism of Senator Lyndon B. Johnson regarding his efforts on behalf of the oil and gas industry. The newspaper article which I have before me indicates that you made the following statement, to-wit:

> I don't know what Lyndon is doing up there in Washington—I want a man to be working for me and not to be a national figure.

Frankly, I do not know of any greater tower of strength that the oil and gas industry has than the Majority Leader of the United States Senate. Lyndon has worked consistently and vigorously to retain the depletion allowance ever since he has been a member of the Congress. Further, since he has been Majority Leader he has used effectively the strength of his dynamic leadership to preserve the depletion allowance and to free the producers of natural gas from price regulation at the wellhead by the Federal Power Commission, all in the face of criticism by many of his colleagues in the Senate from the non-producing oil and gas states.

It is true that Lyndon performs his work in the United States Senate quietly; but, that is to his credit. I subscribe to the point of view that you and I would not be any more qualified to advise him how to conduct his work in the Senate than Lyndon would be to advise us upon the manner of drilling and completing oil and gas wells in Texas.

I must say to you in all good conscience and with the greatest of respect that in my opinion your public criticism of Lyndon and of his record of performance is unfair to him and is calculated to do harm to the oil and gas industry. Frankly, I feel that every oil and gas producer in Texas should give thanks each morning before breakfast that we have Lyndon's support in the Senate and Mr. Sam Rayburn's support in the House on the depletion allowance and the numerous other problems that affect the welfare of the domestic oil and gas industry.

Also, I want to mention that if you are to give full expression of gratitude to Lyndon for favors extended you and your associates, you should recall his interest in your Wise County gas problems before the Federal Power Commission which I called to his attention in my letter of August 31, 1956.

It would be my hope that you would reconsider your attitude toward the Senator and that you would write him a letter correcting the record, not only for your own personal good but for the welfare of the entire oil and gas industry.

With best wishes, I am

Sincerely yours,

Marlin E. Sandlin

August 8, 1958

TO: Mr. Marlin Sandlin
Woodley Petroleum Company
South Coast Life Building
Houston, Texas

Dear Marlin:

I deeply appreciate your letter admonishing me for my perhaps intemperate remarks at a public meeting about our able senior Senator.

I think it should be obvious from the way this thing came about that I had intended no such public statement. I was speaking for no one other than myself, of course, something I want made clear since I was listed in the paper as an official of TIPRO.

In place of writing the Senator a letter, as you suggested, I will forward him a copy of this letter because it will be harder for me to repeat my feelings to him.

Marlin, I am sure you are aware that only a few oil men really devote a great deal of their time and effort to the oil and gas industry. These few along with the various associations have worked diligently to free out industry from FPC control and during the Congress of this year they worked especially hard to incorporate the ideas of the Ikard Amendment into the Reciprocal Trade Agreements Act. It had gained strength in the House, but the administration out-maneuvered the Congressmen in the House of Representatives and with the help of Sam Rayburn and Bob Anderson, this amendment was completely scuttled.

When the Long (Ikard) Amendment reached the Senate, we understood that Lyndon might not be in a good position to co-sponsor it. Since it was the opinion of all the association that this amendment was worth fighting for, we were hopeful that Senator Ralph Yarborough would support it. This he did not do, and it was brought to my attention that the reason for this action was the responsibility of you and Major Parten. I am sure you had your reasons for this, but I am not so sure they should have been allowed to interfere with the program set up by our whole industry group.

I also realize that Lyndon has been a good friend and has worked for the good and the prosperity of our industry. I appreciate personally his help in our Wise County problems in Washington. Speaker Rayburn also has been a good friend but I am sure both he and Lyndon can understand an oil man like me, with so much at stake in the Trade Act, would be critical when I see a possibility of getting out of this mess and no apparent help. I believe that a man of such stature as Lyndon can do about anything he really wants to.

You, together with some of my friends in TIPRO, tell me he does a lot behind the scenes that I do not know about. Of course, I am glad to hear this but I wonder if sometimes he could not go on and admit to the world that we are not such a bad group to be associated with publicly.

Thanks again for your letter.

Sincerely yours,

Johnny Mitchell

July 24, 1959

TO: General Ernest O. Thompson
Judge Olin Culberson
Mr. William J. Murray
Texas Railroad Commission
Tribune Building
Austin, Texas

Dear General, Judge and Bill:

I was disturbed on reading the article written by one of our Houston independent operators on the Port Acres Hearing. I am sure that the three of you realize that I have sincere feelings on the question of wide spacing and am definitely opposed to drilling gas wells on one acre, or less, units. However, I would like you to know that I, in no way, agree with this operator and his attack upon your decision, as I feel that small acre drilling is strictly the fault of the producer and not the Commission. Even though I have expressed myself in this way to the industry, at no time to my knowledge, have I ever felt personally opposed to the decisions handed down by your Commission.

I fully realize that our industry has been operating on Rule 37 for a great many years. If the few individuals who are trying to educate our fellow producers are unsuccessful in getting a majority to feel as we do, then there can be no criticism when the Commission Examiners permit the "1/3rd-2/3rds rule" that is expected in many hearings.

It has been my aim that we, the producers of oil and gas, must decide within our own ranks the differences of conservation. I am extremely critical of any criticism, publicly expressed, to the effect that your decisions were actions of favoritism. Our company has not been granted all of the requests that we have brought forth but, whatever decision was handed down we have accepted, having the same great respect and admiration for you, the Commission. We shall always feel this way regardless of decisions in favor with, or against us.

The main point of this letter is that I shall continue to speak forth my thoughts on wide spacing and conservation but my public statements are intended against an *industry* which I feel needs correcting. I have a deep resentment toward producers who deliberately ruin the reservoirs that nature has provided. It is my truthful opinion that I am personally fighting for a cause that has very little chance of success. However, even though I may not get industry support, the offenders, as far as I am concerned, should be publicly recognized.

If, in your opinion, my appeal to my fellow producers casts any shadows on the three of you and our friendship, please let me know, as I value your friendship far more than a cause for which I can see no true signs of success.

Sincerely yours,

Johnny Mitchell

August 7, 1959

TO: Judge Olin Culberson
Texas Railroad Commission
Austin, Texas 78711

Dear Judge:

Even though I consider myself a staunch friend and an admirer of yours, I am taking a decided issue on a subject that I believe you are dead wrong on. The subject deals with drilling wells on small tracts and your continuing determined stand backing the supposed "Little Fellows" who drill on these small tracts.

Great injustices are permitted to be committed by wealthy concerns which are using what you term the "Little Fellows" as a front man.

Rule 37 on proper well spacing has served the State well in orderly development of our producing field, since 1919. The Railroad Commission has adjusted the drilling distance between both oil and gas wells from time to time in order to improve drainage patterns in oil and gas development.

However, on one hand where the industry's legitimate producers develop their leases under the Commission's requirements of Rule 37, the "Little Fellow" lives under an exception to the rule.

The exception to Rule 37 has created a hunting license for wealthy groups to hunt for small tracts in proven producing reservoirs. These wealthy groups under the guise of being small tract drillers begin their raid on regulated oil and gas reservoirs. These small tracts receive oil and gas allowable that unfairly drains what gas and oil that lies under their small tract in one day's normal allowable schedule, yet they receive an allowable for the life of the well. The "Law of Capture" is making this group richer and ruining reservoirs. The Commission is giving them a "license to steal."

I was amused by your letter dated July 27. The part I enjoyed read, "Dear Johnny, I thoroughly agree with your recent statement that you and others are fighting a losing battle on wide spacing, and you never will win. I do not know of anyone from whom I had rather take a licking on this subject than you my friend . . . but, I ain't gonna take one."

Time will tell who will win the fight on overruling the "Law of Capture" and the subsequent fight on wider spacing which will be the fight on pooling. I must confess that it is as difficult to convince the "Old Guard" of Texas independant as it is to convince you. Most of these successful oil operators came up the hard way, unknowingly of the need of conservation. All they knew in the early days was survival. The "Law of Capture" allowed Derricks to drill wells side by side. The only law in the early days must have been "West of the Pecos."

Today, these old timers, but now leaders, know better but they are a stubborn lot and difficult to turn around.

I don't want you worrying too much about the "Law of Capture" staying as a tool of your Commission to rob his neighbors. The courts will, in the distant future, understand the rights of the many producers and larger land owners.

When that day in court does come, the "Law of Capture" will no longer allow small tract drilling. Pooling will follow to protect the small tract owner allowing him to share only his oil and gas that lies under his small tract.

The "Little Fellow" by that time, will have broken out of his cocoon to be many things; maybe an interstate pipeline subsidiary, maybe an interstate pipeline, or maybe a major independant oil and gas producer, but never an industry leader.

It wouldn't surprise me, Judge, that after your "Little Fellows" get rich enough they will burn up their "Black Robe," buy a "halo," join society, and even go so far as to join our Crusade on Conservation.

Some things they can buy, but to the producers who experienced their reservoirs being robbed by the "Law of Capture," the "Little Man" and their wealthy alter ego will always be included in the independant's "Hall of Shame."

Johnny Mitchell

January 10, 1962

TO: Honorable Joseph E. Swidler, Chairman
Federal Power Commission
Washington, D.C.

Dear Mr. Swidler:

I am fulfilling my promise to you, explaining in more detail what is happening in the interstate and intrastate gas markets here in Texas. The situation has saturated the interstate gas market with "beefed up" reserves with "pay or take" contracts, with "buffer intrastate" lines producing gas on state of Texas regulations which are not only old and chaotic but a legal permit for a "license to steal."

The time has come for the Federal Power Commission, interstate gas pipeline companies, and independent gas producers to meet and explore the serious problems in the inter and intrastate contractual sales of natural gas. Variation in contractual arrangements allowed under current regulations may very well endanger normal growth in gas reserves if something is not done soon. The situation is an explosive one, and independent producers are in no mood to tolerate a continuation of some unethical and perhaps illegal practices.

What really opened the door to formation of interstate pipeline operations was not the financing, planning or presentation before the FPC for permits by the companies, but rather the enormous gas reserves drilled and properly developed by independent oil operators—men who have helped keep this nation strong through their courage in drilling for oil and gas. In the absence of the independent's risk taking, there could never have been enough gas developed to support even one-fourth of the present transmission lines. The financially strong pipeline companies of today which tend to ignore the independent producer should pause and re-evaluate his role in their origin and review their obligation to him.

In the past few years competition among pipeline companies for acquisition of gas reserves has increased significantly. In the beginning of the growth era, reserve purchases were more orderly, and producers were required to develop their reserves according to conservation production-reserve ratio standards in preparation for contractual sale. Proper spacing patterns were observed to assure adequate gas reserves required for EPC certification. Pipelines were required to have 7.5 billion cubic feet in reserves for each one million cubic feet delivered to consumers. On this basis the pipeline company received it's certification and had no trouble acquiring the necessary financing.

For several years each pipeline company complied with the above standards and treated all producers with the same consideration.

But recently the desire to keep up former rates of growth has led some pipelines to launch into some highly questionable, unethical, and possibly illegal practices. Initially, a pipeline purchased all gas produced along its system under one contractual formula that reflected distance and other factors. However, the price of new gas gradually increased as the costs of exploration and development increased, this meant that when a pipeline purchased new reserves along its system, it was faced with an escalation problem involving all of it's original gas contracts.

Consequently, pipelines looked for a method of purchasing new gas that would obviate FPC review and reserve requirements. They found it in short term contracts with small gathering companies on a "pay or take" basis. Since the gatherers were under their own reserve requirements, they promoted and purchased from town lot wells and other wells with limited or questionable reserve holdings—in many instances wells that drain from the very reserves that support FPC approved contracts of other producers.

Since these small tract gatherers are under a "pay or take" contract with intrastate lines, they have no difficulty in acquiring necessary financing, even though their reserve holdings are questionable. This unwholesome development could have arisen only with pipeline blessing and encouragement.

The existence of this dual contract system—frequently in the same producing area throughout this state—obviously works a tremendous hardship on the prudent independent who so carefully met reserve requirements established by FPC. It is authoritatively reported that some 500 million cubic feet per day of unreserved gas is being produced from town lot wells and other systems of unaccounted reserves, thereby eating significantly into the market available to operators with responsible reserve holdings.

The intrastate system in association with small gatherers, jointly through the "Pay or take" method, tie up and purchase town lot wells at a lesser price than the area price paid by the interstate competitor. To compensate for the lesser price, the intrastate pipeline company agrees to take a larger volume of gas each day, exceeding the 1 million per 7½ billion ratio many times. The state laws for gas permit such a discrepancy in the production of gas in Texas. There is no state wide allowable for gas. In most areas which are not under field rules, a gas well is allowed to produce 25% open flow. It is easy to see how town lot wells, not regulated, can drain and ruin any given reservoir.

There may be considerable question whether pipeline practice of contracting for gas with gathering systems without proof of reserves is legal, and it is certainly a common practice in many instances where town lot production is not involved, particularly where residue gas is purchased from a gasoline plant operator at the tailgate of the plant.

While Texas requires wells within a dedicated gas field to have fixed allowables each month on a nomination basis, this requirement is only by individual field. In the absence of a state-wide pattern, therefore, it is possible to have interstate gas sales governed by reserve requirements and intrastate sales limited only by the state's 25% of flow requirement in the same area, thereby causing inequity.

Gas is coming to the same situation that oil has come to. The statutes of the State of Texas should treat gas in the same manner as oil.

Here in Texas, we are heading for a showdown on the "law of capture," the statute which gives the small tract wells not only a permit to drill but a permanent allowable. The most important hurdle will be the "pooling bill." When we eventually win these two battles the small tract drillers won't be around any more.

In the meantime, while we are fighting these chaotic laws, the Federal Power Commission must take a strong stand against certifying the "buffer intrastate gathering system" who feed the interstate line.

Sincerely,

Johnny Mitchell

March 16, 1962

TO: Honorable D. Otis Beasley, Chairman
Oil Import Appeals Board
Interior Building
Washington, D.C.

Dear Chairman Beasley:

I have noticed in the Wednesday edition of *The Oil Daily* that Texaco, Inc. has requested an additional oil import quota, stating the present crude oil stock situation on the West Coast is a potential threat to the national security.

Since no one has apparently challenged the Texaco statement, and since our company is a producer of California crude, I would like to use this means to protest the Texaco representation to your Board.

I protest their statement to you that as principal owner of the Four Corners to Jal, New Mexico, pipe line, Texaco feels it should ship its crude East instead of West as other major producers are doing in the Four Corners area. If this is true, and if their statement that the national security is threatened by the Texaco shortage of crude oil on the West Coast is accepted, then they themselves are threatening the national security in order to maintain a profit position on their New Mexico pipe line.

As a California producer, we recognize that there is a temporary shortage of crude oil on the West Coast, but this shortage has been caused by unrealistic purchasing practices by the major importing companies. These practices have critically reduced the price of California crude to such a point that it is impossible to explore or develop oil today at a replacement cost. We recall that only a few short months ago thousands of barrels of California crude were without any market whatsoever. Today, with the demand for California crude at a premium, the market price for such crude is still far below realistic levels.

We would submit that Texaco, Inc. may, almost alone, correct this so-called threat to the national security by sending at least 15,000 barrels of its Four Corners production per day through the Four Corners-to-California pipe line. If it cannot spare that volume, and does not wish to idle the Jal pipe line, the Jal pipe line flow could be reversed, and the tremendous additional shut-in capacity of Texas and New Mexico be quickly made available to California consumers.

Texaco has been termed a "lone wolf" in the crude oil purchasing business, steadfastly refusing to engage in many, if any, of the normal exchanges of crude oil with other companies. For example, the Texas Railroad Commission has, on numerous occasions, asked Texaco representatives if they would purchase or exchange oil with other purchasing companies if the Railroad Commission set the allowable at a figure lower than Texaco had nominated. A check of the records of this conservation commission will disclose that Texaco, unlike other companies represented in those hearings, will never promise to buy a single barrel of crude oil from other purchasing companies in order to balance their undersupply with the other companies' oversupply.

These matters are respectfully submitted for your consideration.
Yours very truly,

THE JADE OIL COMPANY

Johnny Mitchell
President

April 27, 1962

TO: Honorable Joseph C. Swidler
Chairman
Federal Power Commission
441 G Street, N.W.
Washington, D.C.

Dear Mr. Chairman:

I am taking the liberty to bring up again the same serious problem, concerning recent developments in the F.P.C.'s involvement in issuing certificates to interstate pipeline originating in Texas, whose reserves are questionable.

This unusual procedure is unwarranted because of the assurance you gave me that the Commission's requirements for integrated producers would not be overlooked in the development of interstate lines.

Reserves to develop the long interstate line has become a cloak and dagger situation. I'll again explain the serious situation as best as I can.

The legitimate gas producer who discovers gas reserves, dedicates these reserves to a pending interstate line. This is reviewed by your staff, overseen by the pipeline and delivery of gas is held down to a rate of one million cubic feet per day to each eight billions of reserve.

It is my understanding that some of the long lines have matters now pending before your commission where careful look into the ratio of reserves in place as compared to the actual withdrawals of gas might appropriately be taken by the commission.

A close examination of such subsidiary operations and their interstate lines, the actual reserves dedicated to each such subsidiary line, and the volumes of deliveries made by and through such lines, would reveal, "beefed up" reserves and other facts not dreamed of by you as Chairman but certainly should be aware of by your examiner.

You, as Chairman, would also be surprised to find that numerous corporations with small gathering systems, have been organized and are being operated ostensibly as intrastate lines. When in fact they were always intended for and are engaged almost exclusively in interstate business.

These small gatherers contract with short acreage artist to purchase heavy takes of gas from city lots in the heart of large reserves at prices below the going field price.

Next they represent to the F.P.C. that they have developed greater reserves than are required for their intrastate customer. In some cases they have no intrastate customers. Then they apply to the Federal Power Commission to sell such excess gas in intrastate commerce.

The commission either in good faith or in bed with the gatherers grants such permits and then they carefully set up plans of raiding the large gas reserves owned by large tract operators begins.

It is a strange set of bed fellows, the small gatherers and the long interstate system.

The economic advantage of both pirates are apparent.

In light of previous statements to us by you and members of your commission, we most respectfully await your invitation for a detailing of our view in this grave matter.

Sincerely yours,

Johnny Mitchell

May 2, 1962

TO: Honorable Cecil R. King
House Office Building
Washington, D.C.

Dear Congressman King:

As an independent oil producer in California, I feel it is not out of order for me to attempt to let you know how urgently we feel we need some Congressional mandate concerning the limitations on oil imports.

Whatever the need for liberalizing our foreign trade policy in accordance with the President's Trade Expansion Bill (H.R. 9900), I believe you will agree that certain unique situations require individual attention. You are aware that one of these is petroleum, and that the maintenance of a healthy home oil producing industry is absolutely essential to our nation's security. I believe that the home oil producing industry simply cannot survive intact unless proper restraints are placed on oil imports.

I have no sympathy for any domestic industry which wants protection simply to further the economic objectives of individuals. The domestic producers I know do not advocate any drastic change in the import pattern. We simply want to share in this nation's oil market to the extent that we can hold together a healthy producing segment adequate to meet emergencies.

There has been very little increase in production of oil in the United States since 1956, during which time Venezuela has increased its exports to this country by more than 18 percent, the Middle East by more than 63 percent, and Canada by more than 30 percent. I believe that a small reduction in imports from these countries would injure neither the importing companies nor the source countries.

The State of California has today a seriously depressed oil producing industry. Independents are simply not active in search for the oil which is available to us if more wells could be drilled. That is primarily because of a destruction of incentives by an import policy which is too generous for foreign production at the expense of home producers. In this connection, I am taking the liberty of attaching some statements made concerning this matter recently, including a protest of Texaco's position in favor of higher oil imports into the West Coast area.

We very much feel that unless the national security clause of the proposed Trade Expansion Bill is amended to provide some more positive relationship between imports and domestic production of this defense-vital commodity, the pressures will grow too great to contain imports. The Executive Department has enjoyed ample authority to limit imports since the 1955 extension of the Trade Agreements Act, but has simply proved unable to do so because of the enormous pressures from the State Department and the importing companies.

As a member of the Ways and Means Committee, you are in a much better position than I to know what is possible in this Congress. What I am attempting to say is that we do not wish to be selfish. However, the subject I am attempting to discuss with you clearly pertains to our nation's interest.

As oil producers, we do not have any desire to get in the way of a program which so many believe is necessary to our national well-being in these crucial times. At the same time, we want our problem carefully considered in relation to the Trade Expansion Bill, in the firm belief that it would be not at all inconsistent to strengthen the security clause to the point of assuring a more positive limitation on imports which are found to be involved in our nation's security.

While I am aware that writing a specific commodity amendment into the general Trade Bill is vigorously opposed in significant quarters, that criticism does not apply to all sections of the so-called Steed Bill. You will note that the first section is a non-commodity amendment and does not attempt to tie the President's hands unduly.

May I respectfully urge your personal attention to this matter. If I can be helpful at any time in your important work in this connection, you have but to call on me.

Respectfully yours,

THE JADE OIL COMPANY

Johnny Mitchell
President

September 17, 1962

TO: Honorable Robert S. Kerr
United States Senate
Washington, D.C.

Dear Senator Kerr:

At one time in recent weeks the independent oil producers hopes were buoyed upward by your assurance that you planned to press for moderate new guidelines in the reciprocal trade agreement act to curb excessive imports of foreign oil and their products.

Our hopes were ill founded when we learned from reliable sources that you had no intention of going through with your proposed plan to press for these guidelines in the Senate.

I understand the votes were there in the Senate to approve the amendment on the floor of the Senate. Your leadership could have established a stable ratio between total imports of oil products and domestic production.

When the issue became a decision between hard pressed domestic producers and major international companies who controlled all imports, you and other influential leaders in the Senate were willing to exchange these needed guidelines in the trade act in exchange for some nebulus promises of good intentions by the liberal Kennedy Administration that undoubtedly will never be honored.

I am aware that your own company enjoys a sizeable daily import quota along with the big boys. These quotas are worth thousands of dollars a day to your company. Yet on the other hand you are an elected Senator from the State of Oklahoma where hundreds of domestic producers were looking toward you with the same trust as every independent in America.

Senator, it is fair to say that both the Kennedy Administration and its advisors appear more concerned over which major international oil company will get quotas under the oil import program and who will be in position to influence details of administering it, than they are over assuring this nation of meeting its energy requirements when foreign oil becomes unavailable.

This won't happen in your next few years in the Senate but there will come a day when you wished you would have pressed for moderate guidelines on oil imports. That day will come when unreliable imports of foreign oil becomes a way of life for every American.

Sincerely yours,

Johnny Mitchell

October 2, 1962

TO: The Texas Delegation
House of Representatives
Washington, D.C.

Dear Congressmen:

My deep concern in the threatening growth of unreliable foreign imports prompts me to write. Imports no longer supplement our domestic production but have reached a level where we are actually supplementing imports. Imports get first call, then domestic production is prorated to fill the gap.

For this reason and my deep concern for our energy independence I would like to go on record in charging publicly that President Kennedy is "willing to play politics with the nations' security" in his handling of the oil imports issue in trade legislation.

The domestic independents case against President Kennedy is having his administration teamed with international oil companies to block the Congress adoption of simple guidelines for foreign oil policies.

The Kennedy Administration is willing to play politics with National Security in the import program it inherited.

Domestic groups were defeated in June in their attempt to add an amendment designed to curb the uncontrolled in flow of foreign oil to the Kennedy trade expansion bill.

It was hailed as a politician victory for the President, another demonstration of how naked government power can be used by a team with considerable political know-how and determination.

Actually the victory by the Kennedy Administration was an unnecessary one and likely to be costly and temporary. The Kennedy's team effort in trying to retain power the President does not need, may have committed one of his greatest blunders so far.

A little executive flexibility on this issue would have required the President to do nothing inconsistent with what he proclaims to be his own intent and at the same time would have helped get his powers needed to deal with international trade problems.

The National Security would have been bolstered. The best long range interest of importing companies and source countries served and a measurable gap closed in our international payment.

The domestic producers were seeking a slight modification in the National Security clause in the trade bill. The change provided that in cases where the President found National Security threatened by imports, those imports would be allowed to increase only as domestic production increased.

Such controls would be subject to a broad escape hatch allowing the President to change that ratio when actually necessary to prevent shortages or meet emergencies.

Instead Kennedy leaders chose to deny the Congress any meaningful way to spell out a finding that the present ineffective import program largely ignores the purpose and intent of the law to assure a healthy domestic producing industry.

Executive department officials so far have proved totally unable or unwilling to resist pressure for higher imports. Even after their own findings that imports are excessive to a point of total dependency on foreign crude.

The Kennedy team seems more indecisive than the Eisenhower Administration in handling the import problem.

The present administration in its dedication to the interests of international oil companies, appears willing to sacrifice at the freer trade altar, it allowed concern for monopoly prevention and National Security protection.

Politics has reigned paramount and other objectives have been relegated to secondary considerations.

It now seems clear that this National Security matter is being decided in a political atmosphere. Nothing better illustrates this than the fact political powerful Eastern Congressmen, though fewer in numbers than oil producing states, received special protective concessions on textile and other interest unrelated to our National Security while the most critical problem facing this nation today got only lip service.

When the roof falls in, when the Arabs take control of the world, when America becomes totally dependent on Middle East oil, who takes the blame; not Presidents that time will see out of office, not Secretary of Interiors, like Udall, who we pray will go out with President Kennedy; but the blame will be with you gentlemen who will stay with the ship of state. Most of you will probably be in office when the roof does fall in and the camel drivers from the Middle East become their self appointed dictators of the free world. It will then be too late.

This is why I am addressing this letter to you gentlemen in the Texas Delegation.

Yours very truly,

Johnny Mitchell

October 13, 1962

TO: The Honorable Russell B. Long
United States Senate
Washington, D.C.

I am again taking the liberty to bring up an old subject that has been kicked around in The Senate and The House for the past decade. We independents never asked for much, just a moderate amendment to the "Reciprocal Trade Agreement Act."

A month or two ago we really believed we had such a provision. A new outspoken champion of the domestic independent, Congressman Tom Sneed of Oklahoma, came forward with some blustering attacks against The Kennedy Administration and the major international oil companies for a moderate amendment to the ratio of domestic production with imports of unreliable foreign crude. All of his efforts were thwarted by the combined efforts of adminstration officials ably aided by the international oil companies.

I also know for a certainty that when the House Ways and Means Committee met, there were enough votes to insure a moderate amendment to the Trade Bill, establishing a reasonable ratio between domestic crude and foreign imports. I know, too, that on a certain Friday when it was evident that the domestic oil industry, as well as other commodities, would be insured this ratio, that Chairman Wilbur Mills, who also comes from an oil producing state but has repeatedly allowed his poverty ridden state to be deprived of its natural resources in exchange for favor from the Bureaucratic Administration. He called on under Secretary of Interior John Kelly to lead the way in blocking this amendment and asked John Kelly to have the major international oil companies coerce their Congressmen to vote against this needed amendment. John Kelly, supposedly an independent oil man from New Mexico, supposedly knowledged in the need of the domestic industry and its necessity to our national security, deliberately became the quarter-back for the Administration in blocking this amendment.

This, in my opinion, is the first true break in oil and gas industry relations. This is the first open wedge or deep wound that will never heal in our relations. This is the first time that the major international oil companies have deliberately coerced their Congressmen to vote against the domestic industry.

I was in Dallas soon after a meeting, where 65 oil associations were represented, I told the gathering there that contrary to what they came for, which was to hang Congressman Tom Steed, I personally came to praise him. I told the group that any man so unselfishly devoted to our industry that he risked his future nomination by attacking the major companies in our behalf, deserves my respect and friendship.

Senator, before too many years, imports will be exceeding our domestic production. We've lost the oil wildcatters who kept our nation strong for so many years. Our reserve producing capacity is no longer ours to fall back on. On any given day, at the slightest provocation, the Middle East will close down their wells and issue their proclamation of world supremacy. If you and your colleagues don't believe this is possible, brace yourself, 'cause it's coming.

Yours very truly,

Johnny Mitchell

November 20, 1962

TO: Honorable Joseph C. Swidler
Chairman, Federal Power Commission
441 G Street, N.W.
Washington, D.C.

Dear Mr. Chairman:

We take this informal means of respectfully suggesting that you personally invite our detailed comments concerning recent developments in the FPC evolvement of an area pricing concept for natural gas production. This unusual procedure is warranted, we believe, both because (1) of assurances you gave at our annual meeting last May that interests of non-integrated producers would not be overlooked in the development of an FPC area price concept, and (2) of indications that the industry advisory committee you have named may be unduly dominated by interests contrary to our own.

In particular, we would be pleased to detail for your personal attention our differences with findings and proposals of the FPC staff as stated at the re-opening of Permian Basin area rate proceedings November 15. We doubt that any worthwhile purpose will be served, however, unless we are given by you some reason to believe that our views will not be ignored or relegated to the staff for interpretation and possible distortion.

It is not unreasonable to ask, in view of recent trends, whether the FPC staff predominantly regards area pricing as an opportunity for the Commission to pose as champion of the consumer while in fact promulgating policies tending to promote monopoly practices—which in turn guarantee a kind of federal regulation that is in fact contrary to the interests of consumers and the nation as a whole.

The FPC staff statement appears to recommend what we described during preliminary hearings in the Permian case as the worst possible results which could emerge from the area pricing approach. We take most emphatic objection to the utterly ridiculous implication that some aspects of staff recommendations are based on protecting the small producer. Nothing better illustrates the complete FPC staff distortion of our representations to the Commission than the attempt to justify some below-area prices as a protection to smaller producers. Equally unfair to independent producers is the implication that holding below-area prices in old contracts is affirmed by the doctrine of Mobile (which in fact did not involve independent producers).

These staff findings can be characterized as tending to the rankest kind of arbitrary federal pricing intervention. Discrimination is encouraged wherever short-sighted staff objectives are perceived.

Area pricing in general seems to be viewed as an opportunity to impose arbitrary ceilings which can be lowered without notice or cause, even while refusing to assure such ceilings to all without discrimination.

Without attempting to detail the basis of our concern at this time, we wish to advise that this matter warrants the most vigorous protest of which we are capable. We now feel the situation justifies our stating to you unequivocally that unless and until the views and interests of independent producers are accorded at least some consideration, no further effort will be made to assist the Commission in its area pricing course. You are respectfully advised that the Commission will render a serious disservice both to the industry and to the nation if it fails to redirect the disastrous course of its staff.

In the light of previous statements to us by you and some other members of the Commission, and in view of your unquestioned desire for a regulatory policy which is fair to all interests involved, we most respectfully await your invitation for a detailing of our view in this grave matter.

Respectfully yours,

Johnny Mitchell

February 3, 1963

TO: Honorable John Connally
State of Texas
Austin, Texas

Dear Governor Connally:

As an admirer of yours, I find it somewhat difficult to tell you what you already know. That our state of Texas, which has assumed leadership among all the producing states in the matter of oil and gas conservation, has done nothing to prevent waste and protect the correlative rights where natural gas production is concerned.

Our present day concept of "state conservation" leans heavily on market demand proration. Prorating oil to market demand, despite suspicions of it from within and outside the industry since its inception, has in the main served the best interests of producers and consumers alike. In addition, it has provided a means for developing reserve-productive capacity which proved so essential to our security on several notable occasions. It is soundly based in "physical conservation," but primarily serves as a means for "protecting correlative rights." You recall that our Texas conservation statutes equate "the protection of correlative rights" with the "prevention of physical waste," and market demand proration serves both these objectives.

Moreover, market demand proration for oil has served the consumer with economical prices, because it keeps oil plentiful by encouraging exploration for new fields, even when there is enough oil available to meet current demand. It serves the producer by preventing booms-and-busts which once characterized this business.

Independent producers, the small non-integrated operators, once were most suspicious of market demand proration. Yet they today are having to defend it for oil and fight for it on gas. That's because, under present practice, it is their only hope for getting ratable take that prevents monopoly domination. Market demand proration is the way Texas requires purchasing companies to take ratably as between themselves and all other producers, and keeps pipeline purchasers from unfairly favoring areas and wells in which they happen to be the predominant producers.

This market proration concept for oil, born in Texas in the 1930s, has worked to the benefit of all. That story is a happy one. But, natural gas is another story entirely.

Our gas proration system in Texas is largely a hodge-podge of contradictory and confusing regulations which obscure their real ineffectiveness. In a very real sense they total up to the same kind of chaotic situation which in the East Texas situation of the 30s was considered a "conservation crisis." Wellhead gas prices range without logical basis from 3 or 4 cents to 30 cents per thousand, with wide price differentials occuring in the same field. Even more important, and certainly more directly related to conservation, there is flagrant discrimination in the quantities taken both as between fields in an area and even between wells in the same field. Independents often find it hard to get a market for their gas wells in a field from which a pipeline is drawing from its own or other wells.

A whole range of devices are used to thwart ratable taking of gas. I feel that you know I've long criticized this situation, and called attention to means by which purchasers can favor some wells and producers over others. In particular I have called attention to the mushrooming "buffer gas gathering systems," which often serve as a means of insulating the natural gas transmission companies from their common purchaser's responsibilities to take ratably.

I will refrain from calling any major interstate pipeline companies or gas buffering systems by name; but knowing how quickly you put two and two together, I suspect there will be no doubt as to who I am talking about and what I'm trying to say. That is the urgent need—ratable take proration for natural gas—the absolute necessity for it under present conditions, the fact that it is in the public interest, and the short-sighted irresponsibility of the powerful forces which are opposing it. "Ratable take proration is the only means of conserving this nation's vital natural gas resources." It is the only means of preserving equity among producers and the only means of assuring that all elements of the gas industry survive. "The solution is difficult, but it must be done."

Is there really a surplus of natural gas? The illusion of gas surplus results from the delay of state regulatory policy to require more ratable taking. I am not underestimating the forces of those favoring the status quo, which is, in fact, the present privilege to discriminate in quantities taken as between producers and wells in the same field and as between fields in the same area.

Something must be done about a chaotic natural gas proration system in Texas. The powerful forces opposing ratable taking of gas believe the movement has been killed off. Let them take another look. The present situation cannot go on indefinitely. There are things which can be done and which will be done because they must be done.

When the full facts of these recent developments become better known to the general public and to the industry's continuing practices, I am convinced that gas ratable take proration will be a lot easier to accomplish. As a strong believer in state authority and state responsibility to deal with such problems as natural gas conservation, I believe Texas can work out its problem. But if opposing industry forces are so formidable as to deny corrective state action, then "Washington will take over."

The cause of ratable take proration like state conservation itself may have suffered a serious setback when Bill Murray, Texas' most distinguished engineer and conservationist on oil and gas matters, resigned from the Commission.

Even without him on the Commission the real fight has just begun. And this one I don't believe the independents will lose. If for no other reason that that, losing this fight could mean losing the war, which they must wage for their very survival.

Johnny Mitchell

March 6, 1963

TO: Mr. F.J. Hortig
State Land Commission

I am taking the liberty of answering some of the points brought forth at the hearing last week involving the City of Long Beach tide land development.

The objections were presented by Pauley Petroleum, Shell Oil Co., and Texaco, Humble, Union, Mobil and Shell, who are called the "THUMS" Group. There were several objections presented by this group and with your permission, I would like to answer a few of the objections with the following statements.

Never in my career as an independent oil man, have I ever heard competitive producing companies of such magnitude make such excuse of the word monopoly when the fault of not being oil minded was entirely their own. There is no question in my mind that these companies that are objecting to the City of Long Beach's proposal, had the same opportunities to lease the on-shore leases and the same opportunity to form a combine to bid the off-shore. If a company as small as my company was able to enter the Long Beach area and successfully lease over 1500 town lots since last February, comprising of 300 acres, then I find it ridiculous for anyone to offer a protest.

All companies are aware that an oil field of such magnitude as Long Beach requires unit planning, controlled drilling, pressure maintenance, water injection to prevent subsidence, and most of all, properly drilled wells on a well by well basis to insure maximum economic recovery.

Supposing that we could even consider the case of the objecting companies and were to divide Tract #1 into many parcels, it would require several more operators, joining small units, but it would require not only the drilling of many unnecessary wells, but cause uncontrollable production from each operating unit. It would greatly complicate pressure maintenance and proper injection control, but worst of all, it would create law suit after law suit between unit operators attempting to find a fair equity formula between each unit. I believe that any logical engineer or capable oil man would testify that it is impossible to determine the water levels of the different sands, foresee the fault patterns of this giant reservoir, place the locations for the 1,000 producing and injection wells to be drilled. Only on a planned drilling program under one unit agreement can such a complex operation be carefully carried out. Every well drilled could cause a change of location for the next one. Fault patterns were placed by nature millions of years ago, water levels were also formed by nature and only by drilling can they be truthfully determined. It is impossible to even think that the objectors, all of whom are qualified, could even have the courage to ask that this potentially great reservoir be divided into different units purely for their selfish purposes, completely forgetting that the state and the tax payers will by losing millions of dollars in revenue unless it is kept in one unit and under one operation.

It has been the practice of this industry for years that what a producer can afford, he tackles, and what he cannot afford, he watches. This is my first experience to see major companies object on the grounds of monopoly when their resources are equal to or superior to the operators who are successfully working out of the Long Beach unit. I have come to the conclusion that these producers who are objecting to the Long Beach unit have no objections to the off-shore parcels they control along coastal California. It seems that these objectors only cry monopoly when they, through their own negligence, failed to take care of the golden opportunity in Long Beach.

I am concluding my opinions with the request that this Long Beach reservoir in all its greatness be properly developed into one unit and be preserved as a model field of today and the future. To tear this great field apart for the whims of those that missed the boat would not only be tragic to the State of California but an insult to nature itself.

March 27, 1963

TO: The Honorable Alan Cranston
Controller
State of California

For the second time I object to the statement of Shell Oil Company, Pauley Petroleum Company and Texaco. I feel sure there are other companies of the same stature hiding behind the statement of Pauley Petroleum Company. In my opinion, these objections are being presented to confuse the State Land Commission and other responsible members of the State Legislature.

I find it strange that within this group of opposing companies, and included in the opposition, are State Legislators who are for some reason favoring postponement. They do not realize that this giant field should not become a political football, matching giant against giant and outcome of such fierce opposition will ultimately mean that the State, the City of Long Beach and most of all, the People of California will be the only real losers.

I wish to make one further statement about the opposing combine's statement that they objected to the advance royalty payments being paid by the successful bidder to the State and to the City of Long Beach. I believe I can truthfully state that most of these opposing companies have spent more money either along or in joint operations on foreign shores for foreign oil than the advance bid requested on Long Beach. In addition, the foreign operation of these opposing companies has done far greater damage to the price of our domestic crude here in California than the additional production to be produced in Long Beach will ever accomplish. Even though they are aware that their foreign operations are greatly responsible for the depressing of domestic crude prices, the opposition comes before this Commission to publicly state that to produce oil in the new Long Beach unit by certain combines will seriously handicap the future oil prices in this State and nation.

Gentlemen, again I stand confused for I find it hard to believe that companies supposedly with such large assets as Texaco, Shell Oil Company, Pauley Petroleum Company and their partners can misrepresent facts so broadly, confusing an issue that is so vital to California and to the millions of taxpayers of this great State.

March 29, 1963

TO: The Honorable Edmund G. Brown
Governor
State of California

I am asking for the privilege of enclosing copies of two letters that I have placed in the records of the State Land Commission. Both letters concern the State Land Commission hearings on the Long Beach unit agreement. I am sure that you are aware of what has happened since this unit agreement was approved by the participating companies.

It seems certain to me that an issue of such magnitude and and issue of such importance to the State of California and of most importance, an issue that would benefit the millions of tax payers, would certainly bring top management representation to argue each side of the opinion. Upon entering the auditorium, I immediately sensed that the opposition had already prepared the day's agenda a few weeks ago. The Assistant Attorney General, in an intelligent manner, answered each of Pauley's objections and these answers are also in the records. Credit of respect to the State and the Land Commissioners should be given to Standard Oil of California, Richfield Oil Company and Jade Oil Company, for we came prepared to bring forth our arguments. Proponents of the unit plan came ably staffed by executive vice-presidents of each of the two major companies, along with myself as President of Jade Oil Company. In addition to the vice-presidents, both of the major companies brought a full staff of capable technical men. The City of Long Beach was also represented by their very capable attorneys and technical staff.

In contrast, the opposition had no apparent representation, for each of them was represented by a paid employee of limited responsibility. In my opinion, such representation on such an important issue and before one of your highest governing bodies is disrespectful to the State, the Land Commission and particularly, to yourself. We soon learned that the opposition was fully prepared for the day. They came, they listened, they refused to defend their position, and they left victorious. It was obvious that a delay was agreed upon between the opposition and the State prior to the meeting.

There is no question in my mind that your problems in state government are many and I have no qualms of the integrity of your legislators and officials, but I sincerely believe this is one of the more important issues confronting the State of California today. I urge your office to investigate the delays for I find it impossible and unexplainable that this issue has dragged on since last October and yet to be resolved. Each delay is depriving the State of needed employment, income, use of supplies and materials, and above all, the price objective will be that this giant oil field will make this state independent of foreign oil in the event of a national emergency.

May 10, 1963

TO: Alan Cranston
Chairman
State Lands Commission

Since my letter to Governor Brown of March 29, 1963, two additional hearings have been held by the Commission. At both of these meetings the utter disregard for practical, intelligent industry practices was shown in the arguments presented by Pauley Petroleum, Shell and Texaco. In the interest of the State of California, the City of Long Beach, and the taxpayers who will benefit from the revenues of this large oil reserve, I must write this letter and object to the present actions of the Senate Watchdog Committee headed by Senator Virgil O'Sullivan of Glenn County, California.

I do not have to tell you, the Commissioners or the Governor that Mr. Chapman is the former Secretary of the Interior under President Truman. Also, at that particular time Mr. Ed Pauley was a dominant figure in the Democratic Party. According to my understanding, Mr. Oscar Chapman and Mr. Milton Friedman are members of the same law firm, specializing in matters other than oil and gas. It seems odd that Mr. Chapman and Mr. Friedman would be called in to review the dispute between Pauley Petroleum, Shell and Texaco vs. the City of Long Beach. Under these circumstances and conditions, it would appear to me that Mr. Chapman and his firm should disqualify themselves from this matter due to Mr. Chapman's prior position in the Democratic Party and apparent friendship and connection with Mr. Pauley, one of the participants in these hearings.

The State Lands commission and Senator O'Sullivan and the Senate Committee are aware that the future protection of the City of

Long Beach, the welfare of the State of California and the taxpayers' future compensation in the Long Beach unit is much more valuable to your state than any political consideration that may be involved.

The State Lands Commission and Senator O'Sullivan's Committee have been advised by Mr. Pauley that he wishes a delay on the decision of the State Lands Commission on the Field Contractors Agreement. For this reason I will predict the decision of the legal firm engaged by Senator O'Sullivan's Watchdog Committee. I predict that Mr. Chapman and Mr. Friedman will recommend a delay and further study. Naturally, Senator O'Sullivan's Committee will accept this recommendation and ask for a delay. I further predict that the Governor will urge you to delay your decision and, finally, that your Commission will agree to such a delay and further study. There is no reason for a delay, except that Mr. Pauley wishes it.

May 30, 1963

TO: Alan Cranston, Chairman
State Lands Commission
Sacramento, California

Dear Mr. Chairman:

It would be appreciated if you would have this letter read into the records of your next meeting on the proposed Long Beach Oil Development Program with regard to the Field Contractors Agreement and/or the Unit Operating Agreement of this program. It is my desire that this letter become a matter of record along with my letters of March 6, 1963, March 27, 1963, April 2, 1963 and May 10, 1963, all directed to the State Lands Commission, and my letter to Governor Edmund G. Brown of March 29, 1963.

The recommendations filed by Oscar Chapman May 22, 1963, before the Senate subcommittee investigating the Long Beach Oil Development Program were almost identical to the statements filed by Pauley Petroleum Company and Shell Oil Company in previous State Land Commission hearings. In these previous hearings, Jade Oil and Gas Co. answered all the contentions of Chapman and Pauley in my letters of March 6 and March 27, refusing their allegations.

In engaging the services of Oscar Chapman, I seriously doubt that Senator O'Sullivan had ever previously considered the gentleman from Washington as a logical consultant in this matter. So, it is reasonable to assume that Ed Pauley was responsible for the hiring of Mr. Chapman. Certainly Senator O'Sullivan, a relatively new Senator, having held office only a few years, could not be expected to assume the responsibility of spending $35,000.00 of the State's money to hire outside counsel to review the Field Contractors Agreement and other agreements without the consent of Governor Brown. It is inconceivable that Governor Brown would permit this serious undertaking. The people of California must place the responsibility for the engagement of Chapman's services on Governor Brown and Ed Pauley. Again we ask, is Chapman qualified to review these contracts and render a fair and impartial recommendation? In my letter of May 10, 1963, I said no. In this letter I say, *emphatically, no*! I gave several reasons in my letter of May 10th why Mr. Chapman should disqualify himself in this matter and I feel there are additional reasons why his report should be completely stricken from the records of the State Lands Commission.

Oscar Chapman, aside from being a personal friend of Pauley since the Truman days, is a political lobbyist, not specializing in oil and gas contracts. His best qualification is representing interstate pipeline companies and other public utilities. Aside from being a lobbyist, he has been very influencial in foreign countries by virtue of his previous position as Secretary of the Interior. He has used this past influence to assist foreign investments by American capital, particularly in areas where American business men, without government influence, would find it very difficult to negotiate foreign contracts and agreements. I suspect that Chapman's friendship with Pauley goes deeper than the Truman days, or the era of Governor Brown. I understand that Mr. Chapman assisted Mr. Pauley in his Mexican ventures.

One must wonder if Senator O'Sullivan is aware of Mr. Chapman's performance as Secretary of the Interior. Members of the State Lands Commission, Governor Brown and Ed Pauley are very familiar with the fact that in 1945 the Supreme Court rendered a decision giving title of all the states' tidelands back to the United States Government. Your State of California and all other coastal states were victims of this militant grab by our U.S. Supreme Court. The states affected by this grab fought long and hard in Congress all during the Truman administration in an effort to regain ownership of the tidelands. Mr.

Chapman was a persistant witness against all state's rights to the tidelands. Mr. Chapman was not satisfied to testify just once against the states in their efforts, but appeared time after time in the Congress, fighting vigorously and beyond the call of his office in an effort to prevent states from regaining their tidelands. His determined fight helped keep the tidelands in U.S. hands all through the Truman administration.

Only when Mr. Chapman went out of office with the election of President Eisenhower were the states able to regain their tidelands. It is a strange coincidence that the man who played a key role in keeping the tideland ownership from California should be called in by Governor Brown, Pauley and O'Sullivan to review a contract affecting the same tideland properties, supposedly for the purpose of protecting the State's interests. And isn't it strange what a few will do to change a man's allegiance from national interest to state interest. Now Mr. Chapman has been placed in the position of deciding who will benefit the most, the State (and Pauley) or the City of Long Beach and the people of California.

I am sure that it was not difficult for any of the State Lands Commissioners, Governor Brown, O'Sullivan or Pauley to predict what Chapman's report would read. Paid lobbyists move in only one direction. In all of my experience, never has such an insult been heaped upon the average intelligence of a state's citizenry as in this O'Sullivan-Chapman incident. If Chapman's recommendations are followed, there will never be a Long Beach Unit and the State and City will lose over one billion dollars. The Governor, O'Sullivan and the State Lands Commissioners are able to resist this pressure and reject this report. The Chapman report proposes a devastating abuse to the State and one of our most progressive cities, the City of Long Beach.

What benefit will be gained by Governor Brown, the State Lands Commissioners and O'Sullivan by the State's acceptance of the Chapman report? Political influence so obvious in the Chapman case is sure to hurt the Governor, O'Sullivan and the State Lands Commission. No political reward is worth this sacrifice. No visit to Pauley's Hawaiian Island could be worth it, nor is a trip to his fishing retreat in Mexico worth it.

In conclusion, the greatest insult of the Chapman-O'Sullivan case must be pointed out. The people of California elected in Stanley Mosk, California's Attorney General, one of the most conscientious and capable public servants in California's state government. Aside from the Attorney General himself, his office is ably staffed with most capable assistants. I have had the pleasure of watching them in action at the State Lands Commission hearings on the Long Beach Unit. Unquestionably the Unit Operating Agreement, and other agreements, should have been placed in the hands of the Attorney General. His decision would have been fair because Attorney General Mosk and his staff are qualified. The authorization of the hiring of a known lobbyist, a political friend of Pauley's and Governor Brown's, to decide the fate of the City of Long Beach is an unwarranted act.

Where Governor Brown, Pauley and O'Sullivan, and the State Lands Commission, have welcomed Mr. Chapman with open arms and are paying him a fee of $35,000.00, you can be assured that I feel that any man who fought against states rights as Chapman has in the past should be about as welcome as the boll weevil and the fruit fly.

Respectfully yours,

Johnny Mitchell

July 9, 1963

TO: Mr. R.G. Follis
Chairman of the Board
Standard Oil Company of California

Last week we concluded arguments at the State Land Commission hearings which I hope will lead to a compromise settlement of the dispute as to bidding procedure on the Long Beach tideland.

They have made repeated charges of monopoly and that your company's control of crude as well as crude prices would have a devastating effect on the entire west coast oil industry. In the face of such arguments and my long record of industry statesmanship and industry responsibility, I found myself for the first time in the position that my defense on your behalf had to be the result of a careful, logical analysis of both sides. I had to be sure of the future security of the independent producers in the state of California; their future ability to share the common market of this state and the responsibility of the companies who may win this bid. I also had to be convinced that the winning bidder had to be a deep-rooted California company that would never let this added production affect its responsibility to the small independents.

For the past few years, the California oil industry has become increasingly more dependent upon foreign imports to supply daily requirements. The desire to search for new oil has been severely restricted because of the fact that oil today in California sells for less per barrel than what it costs to find. The small independent refineries who should be searching for California production are instead searching for a license from the Interior Department in Washington to increase their import quota. This is the selfish easy way, the subsidy route. They are not alone, for major integrated companies also constantly appear before this same Import Appeal Board in Washington to increase their import quota.

Historically, some companies in other states have occasionally and others continually held an umbrella over the independent producers' problems. Where in California can we find the umbrella for the California independents. Such a movement cannot begin and end with one man. It must be an industry objective and a company responsibility handed down from one industry leader to the next leader.

We independents with our backs pressed to the wall must look for survival in leadership in someone. To some company whose roots are local and who has not forgotten their industry's obligation and responsibility to the small independent.

Our industry is handicapped with too few great industry leaders and most of all, too few major integrated oil and gas companies with industry understanding. During these past years we have witnessed the burning ambition of most major international oil companies and how success is registered only in the company's growth and profits. The added requirements of these ambitious goals is always passed to each new president of each major oil company as a goal to reach. There is never the opportunity for any new president to stop and consider that maybe it is time that we evaluate our responsibilities to the industry. When has he found time or the courage to say, "What can we, as an industry, do to encourage the independent to stay in business and what contributions, can we make, whether they be the price we pay for oil or other contributions?"

Up to now there has been no hope of industry statesmanship among our many major producers. However, you too have fought the wars of major competition and you have helped build one of the great companies of our era. Secondly, and most important, your company has grown in a state that has seen your fellow independent producers rapidly going out of existence. What greater contribution could you make to the industry than to insure the survival of the California independents. A little less profit on a barrel of California low gravity crude could be your company's contribution to this industry. A few cents a barrel increase in the price of the small independents' crude will automatically mean their survival.

Perhaps the request is too much to ask of one man and one company to accept. But, I firmly believe that if I had the fortune to be the head of a company as large as Standard of California and had my roots imbedded as firmly as your company has in this state, I would devote some of my company's energies to the welfare of this domestic industry.

If I had my choice of walking with kings who control foreign oil, or walking amongst the battered derricks, the leaking tanks, and the worn out equipment of the hard pressed independents of California, I would forego the privilege of walking with kings, until my fellow producers were able to walk alone.

July 10, 1963

TO: Honorable Stewart L. Udall
Secretary of the Interior
Washington, D.C.

Dear Mr. Secretary:

I feel impelled to express grave personal reservations regarding the "technical change" in the oil import program announced June 11.

My objection is not alone to the fact that this change cancelled out a relatively small reduction in crude oil imports east of the Rockies for the six-months period beginning July 1. This scheduled nominal reduction did have special meaning, however, to the hard-pressed domestic oil producing industry—largely because it would have provided a long-awaited symbol of "stability" which the home producing industry badly needs to rebuild its confidence.

Of even greater concern to me than these relatively-few additional barrels of imports permitted by the change is the manner in which the program was amended.

I feel quite certain that you could not have been aware that changing the program in this particular way would directly violate what were believed to be specific assurances from this Administration. Assurances on this precise point, as a matter of fact, were understood to have been given by White House Adviser Myer Feldman, Assistant Secretary John Kelly, Congressman Wilbur Mills, the late Senator Robert Kerr, the Senator Russell Long, all of whom indicated they were reflecting Presidential concurrence.

Last year when the Trade Expansion Bill was before the House, I am advised that Messrs. Feldman and Kelly released to certain members of Congress a "Memorandum for the President," signed by Myer Feldman, in which "intentions" of the Administration with respect to oil imports were summarized.

Mr. Feldman's memorandum of June 25, 1962, (assuming the copies being circulated quote him correctly) includes the following declaration of Administration "intentions":

> "Under the system of import quotas we would establish allocations on the basis of domestic crude production rather than on the basis of estimated demand. This sounds technical, but it is very important to the oil producing states. It will enable the producers and the state regulatory bodies to plan future production."

These plain words were welcomed as evidence that the Administration recognizes the importance of departing from the Eisenhower Administration system of basing allocation on estimates. Yet the change involves a return to an estimate basis with no "adjustment for overestimates or underestimates in Bureau of Mines projections."

In addition to Mr. Feldman's declaration of "intentions," several members of Congress, in explaining their decision not to insist upon stronger language in the security clause stated that the Administration had given them similar assurances for the domestic industry. Senator Long listed the steps which he said the Administration was prepared to take, and declared on the Senate floor that he had "discussed the matter with the President and other responsible persons in the Executive Branch of this Government." One of these points tying "the ratio of oil imports to a definite calendar year" rather than estimates of future domestic production or demand.

On the basis of these widely-published and accepted assurances, members of both the House and Senate declined to insist upon further strengthening of the security clause of the Trade Expansion Bill. I feel that many of these gentlemen will share the conclusion currently being expressed by associations representing non-integrated domestic oil producers. (In this connection a "Liaison Committee," which represents independent producer associations in some 20 oil producing states of the nation, has issued a joint statement contending that this imports change represents "a complete breach of faith" on the part of the Administration.)

I believe that a review of all the circumstances involved shows that this is a most unsatisfactory development, both from the view point of the domestic producing industry and the administration. It is my hope that you will give this matter your personal attention at the earliest practicable moment. The oil imports program as presently constituted, it seems to me, cannot be viewed as satisfactory either by the Administration of the industry.

Sincerely,

Johnny Mitchell

July 11, 1963

TO: The Honorable Wilbur Mills
Chairman, Ways and Means Committee
House of Representatives
Washington, D.C.

FROM: The Honorable John Connally
Governor of Texas
Executive Department
Austin, Texas

Dear Mr. Chairman:

As Governor of Texas, I wish to express in the strongest possible terms my grave concern over the committee's tentative adoption of tax changes adversely affecting the health of this State's leading industry.

In particular, I refer to the proposal that gain from the sale of a mineral property be taxed as ordinary income to the extent that the taxpayer has taken deductions for intangible drilling costs. In my letter of March 23, addressed to you as Chairman of the Ways and

Means Committee, I stressed particular objections to this proposal in these words:

> "This proposal would have a particularly heavy impact *on the small operator* who is attempting to expand and grow by searching for and developing of vital new reserves, then selling them to accumulate dollars to reinvest in new properties with a view to finding and developing new oil. This is a common practice in the oil industry in Texas, a practice which should be encouraged not hampered. Secretary Dillon has testified that this 'recapture' provision would bring in an additional 50 million dollars each year. This burden would fall *on the small producer* because the larger units in the oil industry do not normally sell properties, but rather purchase them. This proposal would reduce selling and reinvestment activities and thus discourage wildcatting for new reserves."

Most non-integrated U.S. oil independent producers are experiencing serious difficulties because of adverse pressures from many sources. It is especially unfair that they—and the defense—vital activity they represent—be singled out for increased tax burdens in this manner.

I personally feel that this amendment would be directly contrary to the President's declared objective of industrial expansion. It will injure that segment of the oil and gas industry that this nation must depend on for the discoveries necessary to maintain proved reserves in proper relation to this country's needs in war or peace.

However great the pressures for so-called tax reform, I submit that no worthy purpose could be served by this particular amendment. The national interest will not be served by any adverse change in the tax position affecting this industry.

On behalf of the ten million people of Texas, and of the oil producing industry which is so vital to their economic well-being, I urgently request this personal assistance of you and the members of the Committee in gaining reconsideration of this matter. I am convinced that the committee's tentative action, if sustained, can do incalculable damage to a defense—vital enterprise and to the most fundamental objectives of the president's tax bill.

Respectfully yours,

John Connally

July 24, 1963

TO: The Honorable Clark Thompson
Representative in Congress
Washington 25, D.C.

Dear Mr. Thompson:

I have returned to California where I will stay until late August. I wish to express to you my sincere appreciation for not only your time, but your sincere efforts and understanding of the problems facing the independents at this present time.

For the first time, I left Washington feeling that through you and your efforts I was able to leave a small image of what we are facing in our fight to survive in this complex international oil industry. Naturally, we independents have the courage to keep fighting and with friends such as yourself, I sincerely feel that we not may, but we will survive these crises.

My very best to you and a thousand thanks from me and this appreciation, I can assure you, comes from the many independents that you haven't had the pleasure of meeting.

Sincerely,

Johnny Mitchell

July 26, 1963

TO: Mr. Johnny Mitchell

FROM: The Honorable Clark W. Thompson
United States House of Representatives
Washington, D.C.

Dear Johnny:

Thanks for your letter of the twenty-fourth.

The most effective job in behalf of the independents was done by your good self in Wilbur Mills office. I did not realize how thoroughly you had sold him until I heard him outline to the committee the typical example which you had given him of the way in which you finance your various projects. Certainly he was very much in accord with your thinking.

We have not taken final action and, of course, anything might happen but I want you to know that as I write this I am very optimistic. Whenever we do act I will get in touch with you immediately.

With kindest regards,

Clark W. Thompson

August 5, 1963

TO: The Honorable John Connally
Governor of the State of Texas
Austin, Texas

Dear Governor Connally:

I have a copy of your letter addressed to Congressman Wilbur Mills dated July 11, 1963, concerning the intangible feature of the present proposed tax bill. Your concern that our intangibles would be taxed as ordinary income equal to our intangible deduction at the time of sale of these minerals prompts me to write.

I have been asked by our mutual friend Robert L. Foree, Sr. and the executive committee to advise you of my experience with Congressman Wilbur Mills in a meeting we had in his chamber with Congressman Clark Thompson who arranged the meeting. Included in the meeting was Robert L. Foree, Sr. of Dallas and Earl Turner of Austin.

I am going to report our conversation and quote Congressman Mills as accurately as I can remember. I don't really believe he will mind my pointing out what a true representative of both the people and private industry he has been during his tenure in office.

In August, 1963, when I was a member of the National Petroleum Council, the most critical thing confronting the independent was this provision of the proposed tax bill. IPAA, Mid-Continent Oil and Gas Association, you, bankers, and many others made strong attempts to take this provision out of the tax bill without any success. On the morning after a National Petroleum Council meeting, I was granted an appointment to see Congressman Wilbur Mills. I had asked for this appointment with the full realization that there was probably no chance to change this provision because I had been advised by some members of the Texas Delegation and others that Chairman Mills was in accord with the terms of this provision and would probably see no reason to alter his decision.

In the first few minutes of our conversation, I realized that Congressman Mills was really unaware of the significance of this provision to the small producer, and for the next hour I explained to him what it meant to the small producers and their investors. I advised him that the small producer was dependent on the small investor for capital for their risk ventures. I asked him if he objected to the intangible cost and depletion and that if he did, there was no point in discussing this provision because we independents first had to have our intangible cost and depletion.

He replied that he thought the oil industry carried great risks and great responsibilities and that an oilman was entitled to the intangibles and depletion.

I proceeded to explain to the Congressman, and he took notes on a scratch pad, that the average small oilman had investors of the $10,000-$20,000 class. On the basis of a $20,000 investment, for example, approximately $14,000 would be intangible costs. The income derived on a $20,000 program, based on a 7-8 year payout in a successful joint operation in many wells, is about $2,500 net for each investor per year. The rule of thumb in a small producer's program is that if an investor wishes to sell his interest, he deducts his income from his investment and subtracts his remaining book cost from the net figure and pays a 25% capital gain on the remaining balance. In further explaining to the Congressman, we took a two year income of $2,500 per year, or a total of $5,000, deducted this from the investment of $20,000, leaving a $15,000 balance. The investor had a $14,000 intangible cost, leaving a book cost of approximately $7,000, which includes depreciation and other items. To figure his tax requirements, the $7,000 had to be deducted from the net figure of $15,000, leaving the investor $8,000 on which to pay a capital gains tax of $2,000.

We then figured this same investment under the new proposed tax law which would require the $14,000 intangibles to be treated as ordinary income. Even though the Congressman had stated that he would advocate reducing the capital gains tax to 19%, he discovered in astonishment that there would be no capital gain to be taken. Immediately he realized that no independent or investor could afford to drill, take his intangible costs as deductions the year of drilling, sell at a later date and then have the same intangibles become ordinary income at the time of such sale.

The Congressman looked at his colleague, Congressman Clark Thompson, who was sitting in on our meeting, and said the Committee had made a grave mistake. They were led to believe that this new proposed feature would be the least harmful to the independent oilman. I suggested that this information must have been furnished by major oil companies who never sell and who would never be affected by this provision. There was no denial by either Congressman, and I did not press the issue.

The dramatic part of this whole episode, as related later by Congressman Thompson, was watching Congressman Mills get up before the Ways and Means Committee, take out the crumpled scratch paper that he had made notations on during our conversation, and advise the Committee that he proposed that this provision be deleted from the tax bill. As you know, the provision was deleted.

Governor, I cannot begin to tell you how fortunate we are to have you as our champion.

Johnny Mitchell

August 20, 1963

TO: Mr. Johnny Mitchell

FROM: Mr. R.L. Foree
First National Bank Building
Dallas, Texas

Dear Johnny:

I want to take this method of trying to say "thank you a million" for your kind and wonderful letter of August 10, 1963, and the previous communications that you have sent me during the year 1962. You have been most generous in your kind words and thoughts. I only hope that I did assist you in making the great success you made as President of TIPRO.

Your active, dedicated, enthusiastic and unselfish efforts continued TIPRO on its ever upward service to the independent oil producers of the United States of America.

Your presentation of actual facts concerning the domestic oil industry to Chairman Wilbur Mills of the Ways and Means Committee of the U.S. House of Representatives was the main act that helped steer the ship of state away from damaging reefs and rocks and in the direction of equalization of domestic producers with international ones. You were in the right place at the right time with the right set of facts that enabled the Branch of the Federal Government regarding the tax bill, and had it not been for you, I feel positive that the domestic oil industry would have sunk lower, and the damage that would have been done, might have been irreparable.

The best always, my good friend.

R.L. FOREE

November 7, 1963

TO: Honorable Dean Rusk
Secretary of State
Washington, D.C.

Dear Mr. Secretary:

It has become obvious that American diplomacy abroad has become subservient to the interest of international oil companies.

It has always been interesting to me to watch how the State Department bristles whenever the interests of American international companies are involved. Theoretically their role is to maintain good international relations for America, which is supposed to explain why they overstate the case for other countries. But I've noticed that when foreign countries clash with American international oil companies, the State Department invariably takes the side of the private companies—at whatever expense to U.S. relations with other countries. What makes it all the tougher on purely domestic companies, in trying to compete with these favored foreign interests, is that it's no longer just our State Department which champions the internationalists causes at our expense. These days both the Interior and Commerce Departments seem to be staffed with people who think they are supposed to act like junior State Departments. A good deal of Secretary Udall and Assistant Secretary Kelly's time is spent abroad these days.

Look at the attitude of our State Department toward the Organization of Petroleum Exporting Countries (OPEC). This is an organization of oil source countries in Venezuela and the Middle East for the purpose of stabilizing the growth of oil production to prevent a crude price collapse. Our State Department is all for similar "commodity pacts" when it comes to tin, wheat, lead, zinc, coffee, and so forth—in all instances the American international companies want them too. But in oil—where the leading American international companies don't want the source countries having anything to say about such matters as how much oil they produce or the price they pay them for it—we find the State Department firmly on the side of the companies and against the countries they supposedly champion.

Now I'm for American exports, just as I'm for unsubsidized foreign investments by American companies—investments which are actually made to fill a need with risk capital of those companies. But when, as now in the case of oil, these companies are investing abroad only because there are American subsidies to be gained, and when they're producing oil in excess of foreign demand and increasing their imports of that excess oil into the U.S., weakening the price of crude worldwide—then I lose my enthusiasm for that kind of internationalism. I say even the companies involved, if they would look up a minute from their selfish pursuit of indirect subsidies, would realize that in the end they're hurting themselves not to mention their own country and the free world.

The U.S. Internal Revenue Service recently announced it was going to examine just why American oil companies post crude oil prices abroad—particularly in the Middle East—which are well in excess of what they actually pay for the oil. Internal Revenue couldn't help wondering how much money the U.S. Treasury is losing because American companies take 27½% depletion on this foreign oil, based on apparently fictitious posted prices. For a long time the statement has been kicked around in international oil circles that "only fools and affiliates pay the posted price for Middle East oil." IRS wanted to know why, if the companies are not actually paying the posted price, they take full depletion on it.

But the State Department is apparently not interested in helping the U.S. Treasury get money when it comes out of the coffers of international oil companies. Just as they are not interested in the side of domestic independents who are injured by the subsidized earnings of international oil companies.

Let's talk about the Argentine situation in which a dozen U.S. oil companies find their investments there threatened by expropriation.

Now I admit I'm prejudiced in favor of American interests anywhere in the world, and I don't like to see American companies kicked around by any high-handed, dictatorial, socialistic despot. But neither am I too happy to see a few selfish companies sitting around expecting to have their chestnuts pulled out of fires they had no business starting. If American companies wish to invest their own capital to develop oil in any country in the world, and are willing to stand the considerable risk of expropriation, then I'm for them. But when they try to take advantage of a U.S. foreign aid situation to develop oil which is not needed in this period of surplus—oil which would not be developed if it were not in part underwritten by our foreign aid and subsidized by our tax favoritism, then I just can't get too concerned over them.

Had it not been for the 600-million-dollars American aid flowing into Argentina, these American companies would never have given a second thought to investing money there to add to the world oil glut. With a sympathetic State Department firmly in their camp, they felt they had financial insurance—Uncle Sam taking the risks, in more ways than one, and them taking a generous profit from this investment.

I would like to bring up that most American foreign aid money goes into the coffers of a relatively few international companies of one kind or another. In some instances our aid dollars never even filter through the hands of the supposed recipient countries at all—but flow almost directly into the hands of U.S. companies. The way I understand is that foreign countries become deeply indebted to American companies. Aid dollars then bail them out—just another way American taxpayers subsidize these American companies abroad. I noticed how our aid dollars to Argentina increased when these oil companies began to get into trouble—to 110 million dollars last year alone.

Again this year, the President's tax cut bill has provisions aimed at the oil industry both at home and abroad. The President's proposals dealing with foreign earnings were made, according to Treasury

Secretary Dillon, because American oil companies under the present arrangement have found it possible to pay virtually no taxes at all on their enormous profits abroad. Furthermore, by adjusting the earnings between domestic and foreign holdings, they actually are able to produce U.S. Oil at less net cost to themselves—an added advantage over their domestic independent competitors.

We are told constantly that domestic and foreign operations must stand together, must fight equally hard for the tax privileges of American oil companies abroad as for our depletion at home. Did you know that other minerals have a lower rate of depletion on foreign production than on domestic? But for oil, American companies abroad get the full 27½% intended for domestic oil only. That would be fine with me if they didn't get in addition a tax credit on their royalty payments, which combined puts them at a tremendous advantage over domestic independents.

Adding to our troubles, on top of all this, international oil companies are privileged to thumb their noses at various national policies intended to limit their activities just as it does others.

Did you know that when America ships foreign aid goods abroad, some 85 percent of it has to be U.S. produced? But when we give away oil as a part of our aid programs, the opposite is true, more than 70 percent is foreign-produced oil by these American International oil companies.

And isn't it interesting that, when other American aid goods are shipped to some country, at least half must go in American-flag tankers. But when oil is shipped under our aid programs, it can go in their foreign-flag tankers at considerable savings to the companies.

Not long ago the President directed all Departments to curtail foreign purchases, and buy U.S.-produced goods for their needs, in order to help in our balance-of-payments problem—but oil was made an exception by the Defense Department. Secretary McNamara, acknowledging that U.S.-produced goods for the military would cost more money, declared nevertheless that "the savings in gold we believe will justify such penalty." But never mind the loss of gold where international oil is concerned.

A few weeks ago John Swearingen, chairman of Indiana Standard, seemed to be explaining the low cost gasoline prices in terms of greater efficiency of refining operation. I challenge Mr. Swearingen to restore domestic crude prices to realistic levels and then see if Indiana Standard can still explain low cost gasoline pricing in terms of efficiency.

Or the kind of example given by Cities Service, whose president, Ed Warren, is our new API leader. We don't need any more crude price cuts—and I suspect the Texas legislative committee will find out Cities Service didn't really need to make those cuts.

The purchasing policies of the leading international oil companies, both at home and abroad, must be changed—in their own interests as well as ours. This blind pursuit of "self-sufficiency" on the part of each integrated company is a dangerous game, and top management must come to recognize it.

I cannot overemphasize the dangers of the present concentration or monopoly trend which is resulting from discrimination against independents at home and abroad.

In the face of this dismal situation another respected friend of the industry, L.F. McCollum of Continental, while serving as API chairman, proclaimed that oil was at last prospering again, and the great need today is for more industry unity.

I suggest that these appeals must first take root in the minds of the major oil company officials. Without some cooperation from the majors who lead the way, without some willingness on their part to forego personal goals which stand in the way of unity, I believe we're heading for the most bitter industry clash in many years.

Let me conclude by saying simply that no company nor no industry segment benefits in the long run by the kind of cut-throat competition which prevails today in our business.

Yours sincerely,

Johnny Mitchell

December 3, 1963

TO: Mr. Jack Rathbone, Chairman
Standard Oil Company of New Jersey
One Rockefeller Plaza
New York, New York

Dear Mr. Rathbone:

I hope you won't object to an old friend questioning a recent statement made by Mr. Emilio Collido, a former undersecretary of the

U.S. State Department and now a Standard Vice-President; defending your Standard's increasing expenditure in developing oil in foreign areas.

He stated: "Standard of New Jersey's export of capital equipment in the development of foreign oil far exceeded our payments for imported oil. Standard was helping our balance of payment."

His statement could ring true only if he closed his eyes to the fact that our nation was now being flooded with unreliable imports of oil. I am sure he didn't check with the present "total" import level of oil and products. For his information it has reached an alarming 29 percent, and imports are on the increase. When Congress first wrote the National Security Clause into the Trade Bill in 1955, total imports bore a ratio to domestic production of 16.6 percent. The intention was to stabilize imports at that level. What are they today? and what of tomorrow? It's time to get concerned.

We are only producing our wells here in Texas eight days each month. Imports are not supplementing our domestic production anymore, but now get first call and then we supplement imports. Independents have reached a point of no return. Small independents who lived day by day on a 30 day allowable are hanging up their tools never to return to the oil fields again. The rest are hanging in there by a thread in the face of an eight day producing schedule and intermittent cut off of their production caused by oversupply of imported oil.

Mr. Collido continues: "Foreign development by Standard and other companies is an important program that would ensure our nation having dependable oil in the future." He knows better, the assumption that we can depend on Middle East, Venezuela and other foreign sources for a reliable source of oil for our future is only for a fool to hang his hat on.

The producing nations of the Middle East have experienced abuse from our State Department and complete domination by our major international oil companies. This domination and abuse will not go on forever until their total oil wealth is dissipated feeding the oil hungry free world. Their day has to come.

The day will come when all the member nations band together in spite of their border differences, turn the valves of every oil well to the left and watch the oil hungry freeworld including our invincible, resourceful United States, fall to its knees, subdued and begging.

I don't know whether you and I will be around when this happens, but the vast resources of energy in the Middle East is destined to play a major role in the everyday life of every nation of the free world.

Our Government, especially our Interior and State Departments and all major international oil companies including your Standard will bear the tragedy of our nation becoming a second rate power.

I don't know where Standard found Emilio Collido but he keeps putting his foot in his mouth deeper as he continues. He stated in defending Foreign Policies:—"Don't slow imports"—"Don't touch the privileged position of the American Companies abroad"—"Keep rewarding and subsidizing the out flow of capital to foreign investments"—"Don't require oil movement in American Tankers"—"Don't interfere with United States Military purchases for foreign oil"—"Just don't do anything to hurt the American Development abroad."

I could go on Mr. Rathbone but I've said enough already. I know of your grass root feeling for the hard pressed independent. I also realize that regardless of your concern, one man and not even one company as large as yours can change the collision course our nation is heading towards.

As I end this letter my deep respect for you is unchanged and I want you to know that the same feeling for you is so evident among all the independents. The independents that have survived this far are still hanging in there but facing unbelievably hard times.

Yours sincerely,

Johnny Mitchell

January 29, 1964

TO: Mr. E. Bruce Street
President, Texas Independent Producers
& Royalty Owners Association
P.O. Box 1110
Graham, Texas

Dear Bruce:

I find it unusual to have to write this letter, but growing troubles in our industry make it necessary that I express my views and those of a great many true independents in the industry.

I believe the prime cause of our troubles is the selfishness and short-sightedness of the major international oil companies. Even with direct appeals from many independents and independent associations, there has been no response of the major companies to review the domestic industry's situation.

Last week the Senate Finance Committee, under an amendment by Senator Williams of Delaware, finally brought forth a needed tax reform on foreign oil operations. As you know, international companies not only get full 27½ depletion on all their foreign operations, but in addition, foreign tax credits—which give them a big advantage over hard-pressed domestic independents and other industries.

The amendment temporarily passed by the Senate Finance Committee would have stopped an oil firm from using any foreign tax credit resulting from percentage depletion to decrease its U.S. tax bill on non-mineral income in another country. When our self-styled industry leaders went into high gear, they persuaded Senator Russell Long of Louisiana (who really knows better) that the whole industry, including independents, oppose this foreign tax credit amendment. In our name Congressmen were told we oppose an amendment which did no more than cut out one minor way international companies gain tax advantage over domestic independents. Working behind Senator Long, who has the reputation as a friend of the domestic industry, they were able to convince three Senators to switch their votes, and soften the amendment to permit use of such tax credits, if they are used against income resulting from oil-related activity, such as refining or marketing. The result is a $10 million cut in tax credits instead of $40 million. No real victory was realized for the important principle of equal tax treatment between domestic and foreign oil investment by American companies.

Last week I was in Los Angeles and it so happened the IPAA was having an annual district meeting. I attended the meeting because I am an IPAA member. The present tax program was discussed by the Washington leaders of IPAA. There was alarm expressed by the leadership of IPAA that the Senate Finance Committee finally developed the courage to adjust the foreign tax credit abuses of major international oil companies.

The IPAA theme was predominantly "let's keep the ship together" and, "Let's all fight together against the tax credit adjustment because if it passes, foreign depletion is next, then domestic depletion."

I could stand just so much of this unity talk and at the end of this rousing "don't rock the ship" theme, I found myself getting up during the question and answer period. I asked this group where industry unity ends and how much the actors and service men, must go broke so the majors can enjoy foreign depletion along with foreign tax credits, and other advantages.

I find no alternative but to be an individual and fight for the rights of my industry, as there is no industry unity today. This has been a one way street for the past eight years and the street is lined with majors, calloused, selfish, and unaffected by the dying domestic industry.

I asked this group why we should fight as a united front for foreign tax credits when we are unable to enjoy such tax credits in America—and this only helps to put international companies in a position to squeeze us out. I told them that it was the responsibility of IPAA and all domestic associations to fight for the rights of the domestic industry, and now is the time for the domestic industry to let our Congress know that this subsidy and double tax credit is entirely unfair both to the domestic industry and to national economy.

I was disturbed when, in the recent Senate Finance Committee hearings, both IPAA and API testified on the tax bill against all advice to let well enough alone. The tax bill passed the House Ways and Means Committee with only the non-aggregation of leases amendment in figuring depletion. This amendment had no effect on independents and probably wouldn't have hurt any major international oil companies that were not misusing it.

Instead of staying home when the Senate took up the bill, IPAA actually testified in a vain effort to reinstate the aggregation of leases. By testifying against all sensible advice, not only did IPAA and API fail to eliminate the amendment on non-aggregation of leases, but they exposed us all to new depletion attacks. Fortunately, the Senate Finance Committee chose only to look further into foreign tax credits abuses by major international oil companies. So the major-organized united industry front finally struck a blow unintentionally for the true independents.

IPAA and API also testified in the House Ways and Means Committee on the tax bill in matters of oil and gas. But if they really think their testimony sold the committee on rejecting the depletion and intangible amendments, they don't know the score.

This victory they risked was won only because key committee members like Congressman Clark W. Thompson of Galveston and Congressman Wilbur Mills, Chairman of the Ways and Means Committee, got for themselves the real story of what this intangible amendment meant to independents. Almost too late key members of the Committee learned of the damage the amendment on the sale of mineral properties would cause the independent producer if he was not allowed capital gains on these sales.

I am not trying to say that the efforts of all the industry in this particular case were not helpful, but you know as well as I that up to that time the major company effort had mainly been to protect their own foreign flanks and protecting our intangible situation was secondary. Mr. Mills didn't learn from them that only the capital gains route would allow the independent producer to enter and remain in the oil industry with investment capital—and that, without a capital gains feature, it would be impossible to secure risk capital.

The Senate Finance Committee has now reinstated most of the major international oil companies' tax advantages. I say we should have the guts to say publicly that the foreign tax credits amendment should stick, and it would be a step in the right direction. Until we get the nerve, good friends like Senator Long will go on thinking they're helping us by working to keep the majors at such a tax advantage over us that we can't hope to survive much longer.

I realize all the reasons you can't be as outspoken as I am now, because last year I had the same responsibility to carry. But this is my opinion—and I believe honest and true independents will understand and agree with me.

It is a sad, sad situation. I hope we decide to stand up and fight before it's too late.

Sincerely,

Johnny Mitchell

February 6, 1964

TO: Mr. J. Ed Warren, Chairman
American Petroleum Institute
c/o Cities Service Oil Company
60 Wall Tower
New York, New York

Dear Ed:

My repeated call for industry unity is not a singular act or belief, for there are many producers who believe as I do. We have made continuous industry appeals to the majors for good marketing practices, better controls of imports, crude price stabilization (only possible by good marketing practices), and for industry responsibility in the purchasing of crude oil by common carrier pipelines owned by major international oil companies. At the present time, major companies are constantly selling their crude lines to independent purchasers and pipelines who immediately not only cut crude prices, but impose pipeline tariffs. The chain reaction of pipeline tariffs and crude price cuts seem to be preplanned by major purchasing companies who are shirking their industry responsibilities.

These are only a few of the things that have caused our hardships. No major company can deny that the independents have tried to cooperate and have appealed to all major companies to show not only industry responsibility, but industry statesmanship. Since these appeals have fallen on deaf ears, there is no alternative left to the independent but to equalize our position with the favored position of the major international oil companies. To do so, we must attempt to eliminate your unfair $500,000,000.00 foreign tax advantage and whittle on your $1,000,000,000.00 foreign depletion.

Ed, are the few cents a barrel that you are cutting back on crude prices because of your poor competitive marketing practices worth it? Are your selfish import requirements and pipeline practices worth risking the loss of your foreign tax advantage and foreign depletion? If the majors took time to look at the situation, I am sure they would come to the realization that it is better to let the domestic producer survive than chance the hundreds of millions of dollars they are destined to lose on foreign tax advantages and foreign depletion.

I realize that many of the majors feel as one major company official once stated, "There are more stockholders in my company than there are independents." This may be true, but the independents have been the backbone of the domestic industry for a great many years and they have not resigned their position yet.

I am not singling you out, Ed, but since you are the Chairman of API this year, you are the logical one for me to write.

Sincerely,

Johnny Mitchell

March 1, 1964

TO: General Louis Pulaski
3530 So. Parkwood
Houston 21, Texas

Dear General:

In our hurried way of life, with our many problems and responsibilities, those that are close to you would be unaware of your illness if we depended on you to let us know. I can appreciate your feelings, for it has always been the nature of the great hearts of God's strong children to conceal their problems so that they do not burden their friends. Much like the lion, king of the jungle, who if mortally wounded, will hide himself away from jungle life and stay secluded until the end. I think this philosophy of the brave and the strong fits you well.

You must certainly be proud of your accomplishments, and your gratitude to God must be bountiful that the men of the Pulaski family have lived in our community for so many years and have accomplished so much in all phases of our society. You have been fortunate that you have been successful in business. You have been fortunate in enjoying a wonderful married life and have reared children and grandchildren that will follow in your footsteps well. You have been fortunate that you have the heart that makes you enjoy giving to your fellow man who is less fortunate than you. Your generosity and compassion for those on the lower rungs of the ladder of life have been a trademark of the Pulaski family for a great many years. For this feeling for your fellow man, I admire you the most.

It has been my privilege since I was a young man to have had the pleasure of knowing three of the finest men that I ever hope to know. You are aware of my deep devotion for your brother, Harry, and your nephew, Eddie. Truthfully, I have never known finer men than you and Harry Pulaski. The same compassion felt by you and Harry was molded in Eddie. Regretably, his life was short and the things he could have done and would have done were shortened because he left this world too soon. Even in leaving at an early age, he set a standard for the young ones to follow.

Naturally I worry about your health, but I feel you have the courage and the determination to continue your many blessings upon mankind that will help you survive this illness. You have the courage, strength and optimism for life that will soon have you up and carrying on your daily acts of benevolence and kindness. Even though you conceal your kindness behind a rough exterior, you have carried on the name of the Pulaski's very well and have become a symbol in this great, thriving metropolis.

We all realize, General, that life is not eternal; that on any given day any of us may depart. The young or the old, or those who have enjoyed as much of life as you. There is no certainty that any of use can live beyond the next sunset. The hope and the joy of living can oft times overcome the strange maladies that ravage our bodies. Very often hope, determination and faith in God can win for us a continuation of life against the greatest odds. I have no realistic evaluation of the chances of your leaving us as there is no true evaluation available from medical science on your condition. Whichever way the pendulum swings, you must know from me today the joy I have derived from knowing you, should we depart without being able to talk together.

I must tell you of the great respect that I have had for you during the many years that we have known each other. You and Harry in a great many fashions chartered my life, helped it financially, established confidence in my future when I was walking the streets as an engineer and, most of all, developed my sense of responsibility to my fellow man. By watching your great contributions in your daily life, I learned the pleasure that comes from giving to those less fortunate than myself. You have certainly earned a well deserved niche in life.

General, I must leave for California again and I pray that you will be up, healthy and waiting for my visit on my return.

My deepest love and respect, always,

Johnny Mitchell

April 7, 1964

TO: Mr. M.J. Rathbone, Chairman
Standard Oil Co. of New Jersey
30 Rockefeller Plaza
New York 20, New York

Dear Mr. Rathbone:

Recently I have taken the liberty of expressing an appeal for some form of industry statesmanship from the major international oil companies. I firmly believe that the only chance for our survival is for the international oil companies to reevaluate their course, and our purpose, in the domestic industry.

Your company, and particularly your subsidiaries, have been a shining example on occasions of holding the independents' hand on many crucial matters. On a few other occasions, your company's policies have caused criticism from the ranks of the independents, myself included.

The competitive fight between the giants of the industry, I feel, compels your company to do many things in the international market that unquestionably affects the domestic producer. I realize it is impossible for one company to stand still and let other international companies pass them by. However, I do believe that in recent months most of the international oil companies are realizing that the small independents' problems are becoming more acute due to excessive imports, crude price cuts, marketing practices and pipeline practices.

You personally have on many occasions been a champion of the independents and I have had the pleasure of listening to your statements that the independents are needed and that we should have the privilege of staying in business along with the large integrated companies.

I have also had the pleasure of several visits with Mr. R.G. Follis and I must tell you that these visits with Mr. Follis have given me new courage and hope that the independents' role in the domestic industry is needed and that we will survive. Such statesmanship from you and Mr. Follis is surely to be a key factor in our survival.

Time is passing, and you have served your industry and your company well. Like all great oil men you unquestionably must step down from your role as the leader of your great company. Before you do, there should be time for you to reflect on the days you first started in this industry, worked your way up through the maze of competition to accomplish the leadership of the giant of all giants. Your good fortune with Standard of New Jersey is only a part of your success, for in your climb to prominence you have had the pleasure of rubbing elbows with thousands of independents who admire and respect you. There is no question that these thousands of friends in the industry ranks look up to you and your company's policies for the umbrella needed for the survival of a great independent industry.

I do not think it is too much to ask of any major company president to look down and extend a helping hand to the hard-pressed independents on the rungs below, for unless he does there is no way that he can approach his responsibility to his industry.

It has been some time since I wrote my first letter to Mr. Follis. Out of this letter I took one excerpt which I had framed.

On my last visit with Mr. Follis, I was pleasantly surprised to walk into his office and see on top of his desk a lone framed object . . . the excerpt which I mention here. He told me it made him reflect that this industry was once made of little people and that little people are still needed, and that he believes this to the point that he will do his best to help the domestic industry survive. I believe Mr. Follis and his principles, and I left his office completely convinced that the policies of Standard of California are unselfish in any phase of competition that will hurt the small independent producer.

If two great men who control two of the greatest companies in the world could follow through with this belief, there is no question in my mind that our survival and our chances to produce and compete will once more become a reality.

As I close this letter I want you to know we are still hanging in there in spite of our derricks being battered, our tanks leaking, and not only is our equipment worn out, but our spirits and hopes as well.

Yours very truly,

Johnny Mitchell

June 10, 1964

TO: Mr. Johnny Mitchell

FROM: Mr. M.J. Rathbone
Chairman of the Board
Standard Oil Company

Dear Johnny:

Probably through some unfortunate delay in the mail, your letter dated April 7th did not reach my desk until May 28th. I mention this because I want you to know that I have not been as slow in replying as the dates might indicate.

Your highly complimentary expression regarding the behavior of Jersey and its domestic affiliates is a source of great satisfaction to me. It is good to know that our managements have conducted themselves in such a fashion as to gain the goodwill and respect of men such as you, who we regard as friends as well as competitors.

Over the years that you and I have known each other, there has been much change in the domestic producing situation. It is inevitable that change will always be with us. Keeping policies abreast of the times is a continuing and difficult responsibility of management. I am hopeful that our performance in the future will be regarded as in the best interests of the industry.

From time to time, I am sure, you are in New York. On one of your trips here I hope you might find it possible to drop in for a visit. Not only would it give us a chance to discuss some matters of mutual interest, but it would give me an opportunity to make you acquainted with some of my associates whom I would like you to know. My present plans are to be in New York most of the time during June and July, and I am hopeful that you might be here sometime during this period. If you would give me a little advance notice by telephone, I am sure we can arrange a mutually satisfactory time to get together.

Sincerely yours,

Jack Rathbone

June 10, 1964

TO: Mr. Fred Hartley, President
Union Oil Company of California
Union Oil Center
Los Angeles, California

Dear Fred:

As a non-integrated independent oilman operating in California, I've reluctantly come to the conclusion that one big reason there is so little room in this state for active independents is that a handful of selfish major companies have a stranglehold on the oil industry out here. Making matters worse, it seems they're using certain so-called "independent" integrated companies like Union Oil of California to beat down any threat to their near-monopoly arrangement.

The bitter fruits of all this already include a depressed West Coast oil industry—one which could easily be flourishing in the otherwise booming economy of California. But what's more important, the long-range cost could include American lives—if we continue to let these selfish interests make California and the whole West Coast dependent upon highly-vulnerable foreign oil. Just because this arrangement is more profitable to importing companies, including Union of California, seems a poor excuse for the present sad state of affairs.

My particular concern right now is with the fact that you allowed Union Oil of California to become the voice of certain selfish major international companies in trying to block a plan to allow oil from the Mid-Continent area to help meet California's growing deficit.

What's more, instead of telling the public the real reasons importing companies oppose making domestic oil available to California, you announced that the proposed reversal of the Texas-New Mexico pipeline to Four Courners and thence to Los Angeles would "impose an economic penalty against California economy of some $36 million a year."

What you are really saying, of course, is that domestic oil would not be as profitable to importers as oil obtained from foreign concessions. It's true that you can get foreign oil to fill this newest 100,000 barrels-a-day deficit about $1 a barrel cheaper. But it is not true that this cheaper foreign oil adds one single cent to the California economy. On the contrary, letting in domestic rather than foreign oil would add immeasurably to the economy of this State, because it would both reduce our dependency upon foreign oil and restore exploratory operations to get our California oil industry going again.

Allowing Mid-Continent states a share of our growing West Coast market might keep you from getting a little more import quotas. And you no doubt like having a federal agency give you handouts in the form of higher import quotas. But isn't it about time to have a little regard for the purpose of mandatory import controls?

If the oil import program has any meaning at all, it is certainly not to empower a federal bureau to assure oil supplies from the most economical source and the most convenient and profitable one to the leading companies. It was rather to avoid American dependency on unreliable foreign oil.

When a proposal was offered to make this possible, i.e., movement of Mid-Continent oil to California through an existing pipeline—you chose to blast the whole idea. To make your opposition seem reasonable you talked about alternatives, all calculated to change the subject.

You know as well as I that, even if Interior reduces imports into Districts I-IV, there still would be no justification for letting imports take the full growth of the District V market. Nor were you serious in dangling the prospect of moving crude and condensates from Louisiana by tanker through the Panama onto the West Coast. You know that, by contrast, the pipeline-reversing proposal is entirely practical—and can be rendered readily economical if Interior officials ever figure out they weren't given power over all imports just to make money for importing interests.

Unfortunately, judging by Assistant Secretary John Kelly's briefing the press after the Washington meeting, you seemed to have provided Interior officials all the excuse they need to buy your bill of goods. They're going to raise the California imports, and do nothing meaningful to encourage movement of Mid-Continent oil to this deficit region. Nor, I predict, will present Interior leaders get serious about cutting imports into the East Coast.

It is a terrible situation when a handful of powerful major companies are allowed to run—and ruin—a defense-vital home oil producing industry. It is even worse when they can enlist the active help of a company which, calling itself an "independent," tries to infiltrate independent producer associations and keep independents from defending themselves. I say this is a disgraceful situation, and that's why I've taken this means of calling your hand. Union Oil and the powerful interests you are fronting have as much to lose by such shortsighted policies as real independents. When the international majors have gobbled up most of us, Union Oil will be next—and federal control is certain to follow soon after.

Yours very truly,

Johnny Mitchell

July 23, 1964

TO: Honorable Ernest O. Thompson
Railroad Commission of Texas
Tribune Building
Austin, Texas

My dear General:

It is my understandable pride and deep admiration for you that prompts me to write this letter. My purpose is primarily to commend you, in the warmest terms at my command, for your long years of unparalleled dedication and service to our industry, our state, and our nation.

When the history of this industry is finally written, you will tower above all others who may be listed among its architects and defenders. You are, in a very real sense, the father and the champion of the state conservation system which accounts for the greatness of this industry.

What makes your achievements all the more noteworthy is that, more often than not, you have had to fight our industry battles without the full support of all industry segments. Indeed, many of your greatest achievements on our behalf have been made over our opposition. The market demand proration system, for example, was viewed with grave suspicion by independents and some major companies alike. Yet it has provided the underpinnings of a stable and healthy industry, and directly made it possible to meet oil requirements of this nation in war and peacetime emergencies.

For years on end, whenever opposition arose to our vital percentage depletion tax treatment, it has been to you we looked for the principal leadership. You have provided not only the intellectual capacity to interpret it in terms meaningful to our critics, but you alone have possessed the personal prestige and the political know-how to win these battles.

Undoubtedly, your most dramatic defense of our industry's good name came during the Suez Crisis when literally the whole world was criticizing our state conservation system on the basis of mythical "oil shortage" created, according to the malicious reports, for the purpose of dictating crude prices. Literally within hours after your dramatic appearance before a Congressional investigating committee, fact replaced rumor—and the correctness of your position became apparent. Those who had helped spread the misinformation, including certain international company officials and their federal echoes, learned the hard way that you would not let this industry be manipulated to their selfish causes.

For you, we devoutly hope there are many fruitful years ahead during which your counsel and guidance of industry affairs will be welcome. Like so many other great industry statesmen, we will not always be able to expect you to continue in your accustomed active role of leadership. I speak, of course, of industry champions and friends of all segments, such as Walter Hallanan, Gene Holman, Jack Rathbone, Gwyn Follis, Morgan Davis, and others—a few of whom have passed on and some of whom are now returning from the front lines. No man can be expected to serve beyond his physical capacity, and this life is not eternal.

As for yourself, General, on some given day you will decide that you have given a full measure to our cause and now is the time that you, like Rathbone, Follis, and others, must turn over the helm of your ship to some younger man. When this time does come, it is my hope that you will help our great Governor select the man to replace you.

What will be needed for our future role in industry history is another General Thompson. The burden of tomorrow's responsibilities demands no less a man. And yours is the opportunity to help select him and perhaps even help guide his footsteps in the difficult first days after he takes office.

Continued leadership is needed to keep state conservation in its proper role, and, as in the past, Texas must provide it. Survival of the kind of industry you have helped mold will require a vigorous champion on the Commission. He must possess the same forthrightness and courage so often demonstrated by you in similar crises. He must be willing, when the occasion demands, to put state strength and responsibility against federal reaction when conditions require an unpopular state course. He must be able to stand up to the giant international companies who would, with no sinister intentions whatever, run roughshod over state conservation principles and small competitors in their logical pursuit of competitive advantage. When, for example, the federal government refuses to restrict imports properly, he must be willing, as you have been, to deal directly with the offending companies at the state level. (We remember vividly how you made importers report import plans to you before you would set their allowables, and how you in years past refused to interpret market demand proration as a convenience primarily to purchasers.) He must never excuse inaction on grounds states must not inject themselves into national issues or into affairs on which federal jurisdiction has been asserted.

General, there's a crisis in this industry today which is in many respects as great as any in its history. Industry statesmanship is failing us badly; no one from industry ranks seems willing or able to provide the necessary leadership to halt a monopoly trend and stave off a threat of complete federal control. Certainly we are being failed also by federal oil officials; there's no semblance of statesmanship among Interior officials in dealing with our import and other problems.

Therefore, we are compelled to look once again to our state conservation body. We will continue to need leadership in the General Thompson tradition. You have always advocated and lived by the rule of reason. One must continue to be governed by the rule of reason. For the future of the independent industry, your consideration in looking ahead to the man that will one day replace you will be another of your great contributions to our industry.

As I end this letter, I would like you to know that my deep respect and admiration for you dates back to the first time I saw you soon after I started in this industry. As the years have passed, this respect for you has continually grown. I sincerely hope that the sands of time allow us to continue to enjoy our good friendship and to continue to work together for this great industry which is so indebted to you.

Sincerely,

Johnny Mitchell

August 16, 1964

TO: Mr. John Swearinger, Chairman
Standard Oil of Indiana
Chicago, Illinois

Dear John:

I wish to thank you for taking your valuable time away from the American Petroleum Institute meeting and discussing the current energy problems we are now facing. The situation as we both see it is extremely grave for energy independence even though government and Congress don't agree.

Today the government attempts to define and limit their roles in the economy. But when the first large-scale consolidation of related companies occurred in the 1870s, regulation did not exist. In the oil industry, for instance, the nation stood by and watched one corporation rise to become a gigantic, multi-faceted operation which exerted its control over almost every phase of the production, manufacture and distribution of this vital product. As it eliminated competition, it became the prototype of the modern corporation. This was the Standard Oil Company. Today, the independent producers are fighting for survival against a dozen such companies, each with the same multi-faceted operations and each working toward the elimination of competition.

The coming of the monopoly and concentration trend has brought almost a complete halt to the major companies' sense of responsibility to the industry. The ambitious selfish goals of these major international oil companies will not permit their industry responsibilities to interfere with the monopoly and concentration trend they are trying to accomplish.

The goal of the major companies seems to be to control as much of the overall industry as they can gobble up, disregarding completely how many small producers, suppliers, drilling contractors, refiners and small service stations are affected. All one has to do today is look at the shuttered doors of all segments of this industry that are going out of business each year. And yet we constantly hear from the major company officials that this is a united industry—that we should combine our thoughts and our efforts and work together, for there is room for both of us. Unfortunately, the independents now walking the streets find this hard to believe.

Major companies cannot believe the fact that they are fostering federal control. It is evident and emminent that if these companies continue to cut crude prices, over-refine, over-produce in their own production states, fight for more imports, produce their gas unratably, fight for foreign tax credits and foreign depletion, import foreign oil with foreign tanker flagships because it is cheaper, and sell their refinery products at unreasonably low prices to squeeze out competition, we will someday see a completely divided industry and federal control.

What will be the future of domestic producers is anybody's guess. The small independent faced with irreplaceable costs will surely fall out like flies. Other independents will hang their tools up and watch this nation run out of oil. Out of the thirty odd thousand producers that gave our nation a reserve surplus back in the 1950s, 10,000 will continue looking for oil and gas but on a limited basis. The bloom is off the flower as far as the independents are concerned.

As far as majors are concerned their prosperity will be bountiful. Imports will make stars out of all you chief executives. Special foreign tax credits will add to this prosperity.

How long will the bloom of imports last? Do we have the time and the strength to reverse this trend of total dependency on foreign imports? With a liberal President like John F. Kennedy, a parks and flower lover like Secretary of Interior Stewart Udall, and a hungry bunch of major internationals gobbling up every barrel of cheap imports assures us that the only way left is down hill and fast.

John, our nation will never again be energy independent. You majors were given too much and too fast. There were great leaders among you majors. The likes of Jack Rathbone, Michael Haider of Jersey, R.G. Follis, Carl Reistle, and I include yourself in this group. There was a great tendency among all you to hopefully keep America strong in energy but the tide of profits from unreliable imports of foreign oil swept all of you off of your feet.

This was once such a great industry. Manned by independents and majors who stood ten feet tall in any crowd.

A few will still be around when the roof falls in. When that happens the same tall men won't be so tall anymore and there won't be any looking back to see how we got off track . . . it will be too late.

Sincerely yours,

Johnny Mitchell

September 9, 1964

TO: Mr. Forrest N. Shumway, President
Signal Oil and Gas Company
Box 17126, Foy Station
Los Angeles, California

Dear Mr. Shumway:

Circumstances have caused me to write directly to you instead of to your department heads, who would normally handle the particular situation to which I am calling your attention. This matter was first referred to Emmet Walters and then to Russ Green, Jr., both of whom I know quite well. The problem at hand goes beyond their jurisdiction, for they are company officials concerned solely with company welfare.

Unlike Signal, a large, integrated, international company, which is able to staff all of its departments with familiar and trusted long-term employees, the Jade Oil Company is a small independent. Consequently, we are often forced to use brokers to acquire our leases. We believe firmly that when we must hire brokers, we deserve the same loyalty from them that we expect from the nucleus of our organization, or indeed that major companies expect from their employees. We consider this to be an important factor in our code of ethics. Fortunately this is the usual case, Mr. Shumway; but regretably, on an issue most important to Jade but rather inconsequential to Signal, your company was instrumental in breaking this code under which we operate. Furthermore, I have been unable to receive any satisfaction from either Russ Green, Jr. or Emmet Walters.

The incident to which I am referring occurred in Dutch Slough Field in Contra Costa County, where both your company and mine have made significant gas strikes. We are both proud of our holdings. You own the Burrough's property as well as other leases. Jade Oil owns the 1,100 acre Capital Company Lease. In addition, we both competed for a few scattered leases totaling approximately 25 acres on the north end of my lease. This lease was assumed to be owned by the Reclamation District. On this basis, Signal made a lease with the Reclamation District for this acreage.

From February through April of this year, after the completion of our first well on the Capital Company Lease, we employed a broker by the name of Art de Golyer to check the entire Dutch Slough area. Under our employment, Mr. de Golyer discovered that the Reclamation Lease that Signal thought they owned was actually owned by the Capital Company. Our Vice President, Frank Mondary, instructed Mr. de Golyer to contact Capital and advise them of the probable ownership of this lease. This he did, contacting Mr. Tom Minney of the Capital Company.

Subsequent to relaying this information, Mr. de Golyer, for his personal benefit, although still under obligation to Jade, inquired of Signal if they were also interested in the lease in question. I informed Signal that they had no right to deal with our broker on this matter. I also talked to Mr. Minney of Capital, who was aware of, and acknowledged the validity of, Jade's position. However, for some unknown reason, he was so bound by the Signal-de Golyer association that he delivered the lease to your company. This occurred after I refused to be blackmailed into meeting Signal's offer. It never occurred to us that the loyalty of Mr. de Golyer to Jade could ever be swayed by any outside influence.

The major part of the responsibility for the persuasion of Mr. de Golyer falls on Signal's shoulders, and the remainder falls on the shoulders of the Capital Company, now known as the Transamerica Development Company. I find it inconceivable that the Capital Company would permit a broker such as de Golyer to convey Jade's lease to anyone but Jade. Yet on this occasion the action of your Land

Department in completely supporting Mr. de Golyer and ignoring the industry code of ethics was instrumental in the Capital Company's action.

In closing, I wish to convey this thought to you. The ownership of this lease in dollars and cents is insignificant to my company. My concern is rather the company loyalty that you expect of your employees and that we expect of ours, whether they be regular employees or those hired from the outside to assist us in our work. When your company interferes with the relationship between the small company and its brokers and feels free to deal with them while they are still under obligation to the small company, then we are traveling a one-way street—all yours.

In California particularly, brokers perform a major function for our company. The single-minded devotion of our brokers is as essential to the small independent as is the loyalty and confidence of your key employees to your organization. We expect your company and other major companies to respect our position in the same way that we respect yours.

I trust that you will give this matter your immediate attention and when you confirm the facts that I have stated will immediately assign to Jade the lease it is entitled to own.

Sincerely,

Johnny Mitchell

September 22, 1964

TO: Jade Oil & Gas Co.
Suite 704 Gibraltar Tower
Beverly Hills, California

FROM: Mr. Forrest N. Shumway
President
Signal Oil and Gas Company

Gentlemen:

We have received a copy of a mimeographed letter dated September 9, 1964, addressed to Mr. Forrest N. Shumway, President, Signal Oil and Gas Company, and written over the facsimile signature of Jade Oil & Gas Co., Johnny Mitchell, President. The letter indicates that copies were sent to other persons and we have been informed that you have distributed additional copies to persons not named therein. The letter contains false statements, attacks our integrity, and in the opinion of our counsel is libelous and is also a slander of title.

You did not inform us of your claimed position until after we had, in good faith and as an innocent purchaser, acquired a lease covering the possible interest of Transamerica Development Company and paid Mr. de Golyer for his services in connection with the acquisition of the lease.

On September 27, 1963, we requested that Transamerica lease to us any interest it might have in the land under discussion. Our moral position in this matter antedates any which you or Mr. de Golyer might have had by a considerable period of time.

We had no knowledge that Mr. de Golyer was acting on your behalf in connection with this transaction. On the contrary, Mr. de Golyer offered the lease to several other oil companies. We have a copy of a letter dated August 21, 1964, which he wrote to Amerada Petroleum Corporation stating that over a week had elapsed since he had submitted the property to Amerada for acquisition and that he was therefore withdrawing his offer. We entered into an agreement with Mr. de Golyer on that same date covering our acquisition of the lease. We had no knowledge of any conflict of interest.

Demand is hereby made that you mitigate the damages we have suffered through your libelous statements and your slander of our title to the lease by a retraction and apology and that this be given the same distribution as your letter to us.

Yours very truly,

Forrest N. Shumway

October 9, 1964

TO: Mr. Forrest N. Shumway, President
Signal Oil and Gas Company
1010 Wilshire Boulevard
Los Angeles, California 90017

Dear Mr. Shumway:

We are in receipt of your letter dated September 22, 1964. Although I realize that your primary concern was to justify the actions of your company, I am sure that you would not knowingly rely on misinformation to support your claims. Accordingly, I feel that perhaps you are not in full possession of the facts.

In the first instance, your executives, Messrs. Emmett Walters and Russ Green, Jr., were aware of Jade's claim on the lease in question. At one time in fact, Mr. de Golyer, acting on instructions from Transamerica with Jade's approval, asked your company if a satisfactory solution could be achieved by partitioning this lease among the respective offset operators. As a result, Amerada, a company whose involvement you specifically cite in your letter, was offered the opportunity to join the lease. However, the proposal came to nought when Signal refused to participate.

On about June 1, 1964, Mr. Frank P. Mondary, our Vice President in charge of Western Operations, and I met with representatives of Humble, Gulf, Union of California and Signal in Humble's offices. The purpose of the meeting was to discuss the directional drilling of our #3 well offsetting Union's leases. During the conference, Mr. Mondary called your attention to the fact that Mr. de Golyer was working for Jade in attempting to acquire a lease on the 25 acres in question. Your representatives at that meeting refused to acknowledge the validity of our position. If you have chosen to rely upon your own interpretation of Mr. de Golyer's status, together with any representations he made, that will not preclude us from showing that we were in fact paying Mr. de Golyer and were paying him for the purpose of checking properties in the Dutch Slough area and in obtaining leases for Jade, and we have paid him compensation for services in conjunction with work which he did in this area.

After June 1, 1964, I called Russ Green, Jr., Emmett Walters and Don Wagner, but their uncooperative attitude made it impossible to reach a solution. At that time I made it known to them that Jade was claiming the right to this lease because Mr. de Golyer was working for Jade. I must concede that quite often companies other than Signal have, perhaps inadvertently, infringed on the rights of others. However, the executives of responsible companies, when informed of the situation, are generally quick to rectify such errors.

Before your company acquired this lease in August, 1964, I called both Mr. Minney of Transamerica and Mr. Walters of Signal and stated that if Signal purchased the lease directly from Capital Company without paying brokerage fees to Mr. de Golyer, I would have no right to object since Capital Company, in fact, did have the right to assign the lease to anyone it pleased. I further stated, however, that if Mr. de Golyer was paid, then Signal was negotiating with fiduciary agent and infringing upon my rights. This conversation, in fact, occurred before Signal obtained this lease, which was not on August 21 as stated in paragraph 4 of your letter of September 22, but after August 27, which is the correct date of the execution of the lease by Captial Company although the lease bears the date of August 15, and was delivered to you by mail after August 27. We believe from the information we possess in our investigation of this matter that Signal employees did have knowledge that Mr. de Golyer was working for Jade prior to the time Signal obtained this lease and that we will be able to establish that fact.

Your position of this issue is not warranted as you seem to think because you requested in September, 1963, that Transamerica lease to Signal any interest it might have in the land under discussion. In August, 1962, we purchased from Transamerica a base lease of 1100 acres, with the understanding that this was all the available land. As routine protection in acquiring a lease as large as this, it is industry practice to assume that after abstract and title define the leasehold limits, the size of the lease may be either more or less than the originally stated acreage. I was advised by Mr. Tom Minney of Transamerica that if he had known of this 25 acre tract, it would have been a part of our original lease. Consequently, Jade's claim antedates Signal's by more than a year.

We have reviewed this matter with counsel and have concluded that our position is just and correct, and that if Signal will not agree to transfer of this lease to Jade, we are prepared to file a law suit against your company in order to protect our rights. Your contention that you have been libeled is incorrect, and I might point out that I have been advised by counsel that truth is always a defense in a libel action. I have fought too long for principles to allow your company to abuse my company and the rights of independents. I know that unfortunately most independents, because of their size, have to accept, without recourse, the acts and dictates of major companies. On this occasion, we are prepared to litigate the issue and to defend our position before a jury in the proper court.

Yours very truly,

Johnny Mitchell

November 4, 1964

TO: Jade Oil and Gas Company
Suite 707 Gibraltar Tower
Beverly Hills, California

FROM: Mr. Forrest N. Shumway
President
Signal Oil and Gas Company

Gentlemen:

Since my letter of September 22, and also as a result of your letter of October 9, 1964, to which this letter is a reply, a detailed investigation was made of the facts relative to Signal's acquisition of the 25 acre Dutch Slough lease from Transamerica Development Company through Mr. Arthur de Golyer, whom you have alleged was your agent in acquiring that lease for your company.

Our thorough investigation has included, among other things, a complete interview with Mr. Arthur de Golyer and with all of the Signal personnel involved in this matter. Statements have also been taken from persons of other companies who attended the June 1, 1964 meeting, to which you refer in your October 9, 1964 letter. Also, a meeting was held in Oakland with representatives of the Transamerica Development Company. All of the statements of the impartial witnesses and all of the evidence gathered, together with the facts discovered by our investigation, fully and completely support our conclusion that the inferences and direct statements contained in your letters of September 9, 1964, and October 9, 1964, are erroneous and without foundation, and we remain convinced that there was no improper conduct by Signal whatsoever in the acquisition of the 25 acre lease.

This investigation strengthens and affirms the position outlined in my letter of September 22, 1964, that the publication of your letters are libelous in nature and further constitute a slander of our title to the 25 acre lease obtained from Transamerica Development Company.

I am, therefore, again demanding that you publicly retract the libelous statements published through your letters of September 9 and October 9 with as much force and coverage as given the original statements.

Yours very truly,

Forrest N. Shumway

November 13, 1964

TO: Mr. Ralph W. Trueblood, Jr.
Vice President, Belridge Oil Company
President, Oil Producers Agency of California
701 National Oil Building
Los Angeles 17, California

Dear Ralph:

As a member of Oil Producers Agency of California, and an independent oilman trying to make a better future for my company in California, I think it's my duty to protest the way an organization supposedly representing independent producers is being misused to our detriment.

I'm particularly incensed at the way our executive vice president, Stark Fox, publicly condemned one of our greatest industry statesmen, Jack Rathbone of Jersey Standard. I also must protest his widely circulated memorandum of September 14 claiming that District V is not becoming dependent upon unreliable foreign oil, and that domestic producers are sharing properly in West Coast market growth.

The press quotes Stark as accusing Rathbone of having "aligned himself" with "domestic producer groups east of the Rockies (who) have launched a concerted drive to bring District V into the nationwide import program." He adds:

> "They have done this despite the finding of a distinguished Cabinet committee that conditions in District V are so different that they require differential treatment in the best interest of the import program and, hence, national security."

His view was echoed by the executive vice president of the Independent Refiners Association of California, James Demaras. He called Rathbone's speech before Texas Mid-Continent Oil & Gas Association "without question, the most naive and irresponsible I have ever heard from an industry leader of his stature, adding:

> "Rathbone's statement that District V should be forced into a single common nationwide district so far as petroleum imports are concerned has, like those of other individuals and groups east of the Rocky Mountains, completely ignored the most elementary and basic economic facts that prevail in this district.

Now I say that the people who pay Demaras' salary may, for selfish and short-sighted reason, want him to protect their sources of more profitable foreign oil. But for the life of me I can't understand why we independent producers, who Fox is supposed to be representing, shouldn't protest things he's doing to keep imports high—so high that our California crude prices have been cut from $3.05 when import controls began down to $1.90 today.

I don't know whether you realize it or not, but from the earliest days of the voluntary and the mandatory import program, Fox has been preaching the doctrine of isolating the West Coast. The fruits of this policy include a serious depression for independents in California. Only a handful of independent producers are still able to even search for oil out here. He succeeded in convincing the finest import administrator (now a Sinclair vice president), that California would somehow benefit by letting importers take almost the whole West Coast market growth. The result has been that since the Mandatory Program began in 1959, District V imports are up around 150,000 barrels a day while production in this District increased only 3,000 barrels a day. And the oil we do produce here, we have to sell at distress prices—on terms that suit the importing companies who are also the purchasers of independents' oil here.

In his published memorandum dated September 14, Fox claimed that since most of the increase in imports come from Canada, there's nothing to worry about. He says "the Federal Government considers Canadian oil as domestic oil insofar as national security and imports policy are concerned."

I say that no government official believes that, Demaras doesn't believe that, you don't believe that, and Stark Fox knows it isn't true.

In the first place, Canada itself is a big importer of oil. They import about as much cheap foreign oil on their east coast as they are forced to dump on us from their surplus producing western provinces. As a matter of fact, the same international companies who import here have it rigged to guarantee Canada that any time they increase their imports of Venezuelan or Middle East oil into Canada, they'll dump an equivalent amount of Canadian oil on the U.S. market.

It is true that Canada was given an exemption from import quota controls on grounds that this imported oil is safer because it moves by pipelines from Canada. But in an emergency, Canada would need all her own oil and more, especially if her imports were cut off or cut down.

Even if Canadian oil were as reliable as domestic oil, is that a sound reason to keep letting Canadian imports grow so much faster than domestic production. Mr. Fox is taking the ridiculous position, in effect, that Canadian oil is more secure than U.S. oil.

If Mr. Rathbone is "naive and irresponsible," as these California "independent" spokesmen claim, then so are a lot of other people. In fact the whole idea of an import program, judged by their standards (of what's good economics for importing interest) is equally "naive and irresponsible." And the ranks of the "naive and irresponsible" seems to be growing.

I refer you to a memorandum dated September 4, 1964, just 10 days before Stark Fox's helpful memorandum, issued by Richard C. Sauer, Executive Vice President of the Standard Oil Company (Ohio). Ohio Standard shows just how "naive and irresponsible" is the whole idea that District V should be set aside to sop up import surpluses. They contend that "the oil import program should be based on a single country-wide quota system," adding:

> "The present oil import program has been in effect for a number of years, and even in its origin, like Topsy, 'just growed'. It has been workable in a legal sense, but administrative problems have been followed by patchwork adaptations until the present confused structure has developed."

Ohio Standard illustrates how ridiculous the idea of separating District V, on grounds of oil production deficiency, by pointing out that District I doesn't get this importer-favored status "even though it is farther removed from the prolific domestic producing area than is Southern California and is closer to the principal sources of foreign supply."

This newest addition to the ranks of the "naive and irresponsible" shows how, through this separate application, "the normal flow of United States crude oil is artificially blocked with harmful economic consequences," how "this favored treatment for the West Coast discriminates against producers and refiners in the rest of the country," and how "our national defense is significantly harmed."

Supposedly representing independent producers, Oil Producers Agency of California (OPAC) spokesmen have worked with certain key government officials—and inside independent groups—to block every move to link this whole nation under a single import system that makes sense.

Since the price here fell so low, Stark has been armed with the contention that it would be "uneconomic" to bring Texas-New Mexico oil into California. Ignoring how crude prices here in California got so low, and how this itself should be a national security concern, I'd like to quote a paragraph from the Ohio Standard Brief:

> "Furthermore, even if District V crude were priced lower than West Texas crude, is this an adequate argument for special treatment? West Texas crude will still deliver to Los Angeles for less than it will deliver to Philadelphia, assuming in both cases a modern big inch pipeline. There is no valid argument for this artificial line of demarcation . . . "

Jersey Standard officials, including Jack Rathbone, have taken the same position. And they're not looking for a federal subsidy, which Fox accuses independent producers in Texas of wanting when they mention domestic oil should be shared in West Coast market growth.

Since I operate in both Texas and California, I have been accused of selfishly wanting to find a California market for my Texas oil. This is silly. My interest is far broader. My interest is in seeing that domestic producers in every part of this nation get a fair break. When we shut off one area for cheaper foreign oil to dominate, producers there and everywhere else in America suffer. The crude price depression in California, a direct result of this importer-inspired arrangement, hurts crude prices everywhere—but most of all the people Stark Fox is supposed to represent. I'm not being either pro-Texas or pro-California in saying that, if the import program has any purpose at all, it must be changed to bring California back into the Union. It just won't work this way; that's a fact which should be obvious to everybody by now.

Since Mr. Rathbone's position has been called "naive and irresponsible," I think it's time to bring out into the open some OPAC actions which I think really fit that characterization.

I'm talking about efforts to confuse an easily-confused Interior hierarchy with the argument that California production will make a comeback without doing anything about the separate-treatment system.

Secretary Udall recently confided to some astonished oilmen that California independent producer spokesmen had assured him California would be self-sufficient in four or five years. It seems he had been told in confidence about "a new breakthrough" in a secondary recovery technique. While this supposedly couldn't be discussed publicly for company economic reasons, it was described as so significant that California wouldn't need imports in four or five years.

Mr. Udall apparently took this irresponsible "information" at face value. Maybe we shouldn't expect him to know better; he hasn't taken the time and doesn't have background enough to understand Interior responsibilities pertaining to oil. But I say to you that Assistant Secretary Kelly knows better, as you know better, and every other responsible oilman knows better, and I can only guess why Kelly hasn't seen fit to give Udall more responsible information on which to base Interior policy.

Of course there's progress being made, rather significant progress, in secondary recovery techniques. Your company and my own are among those who stand to benefit by it through the years. But even the most optimistic predictions imaginable couldn't possibly support the Udall notion: that California will soon be self-sufficient, so there's no need to stop letting importers have a free rein in California.

Mr. Fox's memorandum tried to leave the impression that, without prejudice, he was simply trying to set forth "the facts." I'm interested in facts, too, but I'm not going to pretend to you or anybody else that I'm impartial. I'm damned partial—to domestic independent oilmen, as I think Stark should be. So I want to see changes which will let us independents stay in this great business. And I'm too tired of watching the independent lose out to be very diplomatic in speaking my mind.

For a long time, I've wondered and worried about the Oil Producers Agency of California's position favoring separate treatment for District V. So long as this idea of bringing California back into the nation could be blamed entirely on Texas independents, I thought it best not to protest. Now that some of our major company statesmen have been condemned for stating an obvious need, I felt I had to speak up.

With the kind of representation California independents are getting through OPAC, it's no wonder domestic independents find themselves in an economic strait jacket. It's no wonder we're not drilling enough wells even to find enough oil to replace the oil we are producing.

The program's got to change, even if it is a lot more profitable for a few to keep the present arrangement. OPAC has a fine opportunity to help bring about responsible and orderly change. I'd like to see you and other intelligent independents start making OPAC decisions, and not have our decisions made under the table by those with a selfish interest in keeping the present setup.

Yours very truly,

Johnny Mitchell

June 9, 1965

TO: Honorable Nicholas deB. Katzenbach
Attorney General of the United States
Washington, D.C.

Dear Mr. Attorney General:

Speaking as a private citizen of this great nation and as an independent producer of domestic petroleum, I believe the time has come for a comprehensive investigation of Interior Department's handling of the oil import quota allocation program and related matters.

I am convinced that an early airing of Interior's handling of oil-related matters will prove in the long-range best interest of this industry as well as this Administration.

Without detailing the many grounds for suspicion, or pretending personal knowledge of overt acts of wrongdoing, I can tell you that there are enough surface indications to justify serious concern within the industry and the Administration. A major company official, for instance, is reliably quoted to the effect that there are things going on which could ultimately prove extremely embarrassing both to the industry and the Administration." I believe this can best be avoided by a Justice Department inquiry in advance of any Congressional or press media exposure.

In particular, I suggest that attention be focused at this time upon negotiations through which Interior officials committed themselves to a special arrangement by which Phillips Petroleum Company may import some 50,000 barrels a day of products via Puerto Rico, without having to pay the normal excise tax upon imported products of this kind. This privilege involving a $600,000,000 petrochemical complex, was awarded this company by negotiation with Interior officials, without the benefit of competitive bidding. This means that product imports of other competitive companies will either have to be cut back to make room, or total non-residential product imports will increase. Also it will require Presidential sanction, since the Executive proclamation will have to be amended to allow for it. I feel sure the President will wish to consider this particular Interior deal more carefully because it does require his direct affirmation.

It is hard to believe Secretary Udall's reasoning that the private negotiations with Phillips in Puerto Rico was done in behalf of Puerto

Rico's economy. Any reasonable man knows that Phillips, Standard, or any other major international company, is interested in the earning power for itself and its stockholders. We have witnessed the selfish ambitions of all these major companies and you can rest assured that Phillips' move is motivated to gain an advantage only for itself. I further believe that Secretary Udall does not have the basic background of oil and gas matters, nor the courage to grant such a permit which will unquestionably destroy the import program without advice from some other authority.

This special handout to Phillips will automatically ensure that every major international company will soon apply for this same privilege that has been granted Phillips Petroleum Company.

Only recently, when the Phillips negotiations were announced, Jersey Standard objected to Secretary Udall's and Phillips' private agreement and when it became obvious to Standard that the cards were stacked, Jersey Standard immediately changed its tune and asked for an import quota into Puerto Rico. Even more recently, Indiana Standard applied for an import quota of its own. The Phillips handout will unquestionably open the flood gates of uncontrollable imports and this present move by the Interior Department will not only criple the import program, but will ensure added hardship on the domestic oil industry.

It should be recalled that in 1961, the same company negotiated helium contracts with the same Interior officials, without benefit of competitive bids. The General Accounting Office later charged that this deal amounted to a $155 million government giveaway. The helium contracts also were awarded by negotiation. The Secretary later admitted that Interior had "made a mistake" in negotiating the contract without insisting upon renegotiation clauses so the government could recover funds over and above "an equitable return."

It has always been my understanding that it was your Executive responsibility to prevent unfair competition. On several occasions officials of the Justice Department have expressed concern with the monopoly trend in this industry, and with the tendency of the Oil Import Program to aggravate that trend. Representatives of your Department were reportedly active in developing President Kennedy's Petroleum Study Committee report in which it is declared that the present quota allocation system "places the federal government in the role of distributing a considerable economic advantage," adding: "The distribution of an economic advantage of this magnitude is bound to bear upon the financial return of individual companies, the structure of the industry, the functioning of the economy, and other national objectives." This is certainly happening, and there should be at least some evidence that one Executive Department is concerned.

Again let me state that I am personally aware of no misdeeds involving a federal official. I am aware, however, of a growing feeling within this industry that there are unwholesome developments which could prove seriously embarrassing to us all. Neither the industry nor the Administration needs another Teapot Dome scandal.

Accordingly, I believe the best approach is the most direct one, and that the Attorney General should let the industry and the public know that it is inquiring into and making itself aware of anything which could be improper or inappropriate.

Your personal attention to this matter is urgently requested.

Respectfully yours,

Johnny Mitchell

June 17, 1965

TO: Mr. Johnny Mitchell

FROM: Mr. Robert L. Wright
Acting Assistant Attorney General
Antitrust Division

BY: Mr. Joseph J. Saunders
Chief
Public Counsel and Legislative Section
United States Department of Justice

Dear Mr. Mitchell:

The Attorney General has asked that I acknowledge your letter of June 9, with reference to oil import quotas to Phillips Petroleum Company.

Thank you for your expression of interest in this matter. As you know, administration of the oil import program is primarily the responsibility of the Department of the Interior. However, you may be assured that your views will be carefully considered in connection with any action the Department may be called upon to take.

Sincerely yours,

Robert L. Wright
BY: Joseph J. Saunders (signed)

July 15, 1965

TO: Mr. Johnny Mitchell

FROM: Mr. Stanley Learned
President
Phillips Petroleum Company

Dear Mr. Mitchell:

Attached is a copy of letter written today to the Honorable Nicholas deB. Katzenbach, Attorney General of the United States, which is self-explanatory.

It is evident to me that your letter of June 9, 1965, to Mr. Katzenbach was written by you for the purpose of, and with the malicious intent of, doing damage to Phillips Petroleum Company. We deem your inaccurate statements and the press reporting which your letter inspired as harmful to Phillips. For this reason I wish to clear up any misunderstanding by sending a copy of my letter to each of the parties to whom you mailed your letter. Will you please, therefore, be kind enough to send me your list with addresses?

Very truly yours,

Stanley Learned

July 15, 1965

TO: The Honorable Nicholas deB. Katzenbach
Attorney General of the United States
Department of Justice
Constitution Ave., -Ninth to Tenth Streets
Washington, D.C.

FROM: Mr. Stanley Learned
President
Phillips Petroleum Company

My dear Mr. Attorney General:

In a letter to you of June 9, 1965, Mr. Johnny Mitchell of Jade Oil and Gas Company made several statements and suggestions reflecting adversely upon officials of the Department of the Interior and Phillips Petroleum Company and requested that your Department undertake an inquiry into Interior's administration of oil-related matters. Only recently a copy of this letter was brought to my attention. While it is not necessary to respond to all of the inaccuracies in the letter, I feel it is my duty, on behalf of Phillips Petroleum Company, to cover certain points.

1—Mr. Mitchell's letter states:

> "In particular, I suggest that attention be focused at this time upon negotiations through which Interior officials committed themselves to a special arrangement by which Phillips Petroleum Company may import some 50,000 barrels a day of products via Puerto Rico, without having to pay the normal excise tax upon imported products of this kind. This privilege involving a $600,000,000 petrochemical complex, was awarded this company by negotiation with Interior officials, without the benefit of competitive bidding."

I can assure you categorically that there was no arrangement, special or otherwise, made or even considered with officials of the Department of the Interior concerning exceptions of any kind involving taxes, nor was Phillips accorded preferential treatment over other companies in connection with any aspect of the project. Terms and conditions upon which the Puerto Rican petrochemical complex would be established were made public in a release of the Department of the Interior of May 14, 1965, which called for the Government of the Commonwealth of Puerto Rico to enter into a binding agreement with Phillips based upon such terms and conditions set forth by the Department. The agreement between the Commonwealth Government and Phillips is a matter of public record. These documents make clear that the Department committed itself to no tax exceptions as alleged.

It is pertinent to point out procedures followed by the Department of Interior prior to its determination to grant an allocation for the Puerto Rican petrochemical project. Beginning in 1961, the Government of the Commonwealth of Puerto Rico considered that the historic limitation placed on importation of petroleum into Puerto Rico whereby imports were limited to the two companies existing in 1958 was unsound public policy for the reason that artificial restriction was placed upon the Island's propsects of economic growth. Thereupon, the Commonwealth Government initiated discussions with the Department and also with a number of oil and chemical companies in the hope of eliciting from such companies proposals for a petrochemicals-oriented refinery. Several proposals were received by the Commonwealth Government but no proposal was deemed acceptable to the Commonwealth until that submitted by Phillips Petroleum Company in December, 1963. Thereafter the Commonwealth Government communicated with officials of the Department and requested that an allocation be granted to Phillips for a petrochemical project. By release of June 25, 1964, the Department of the Interior announced that a public hearing would be held in Washington, D.C. on July 31, 1964, to receive testimony on all phases of the mandatory oil import program as it related to Puerto Rico. The release stated further that "The hearing is being called as a result of representations from the Economic Development Administration of the Commonwealth of Puerto Rico pointing out the unfavorable economic outlook for Puerto Rico and the stimulus to continued economic growth in the Commonwealth that could be provided by establishment of a petrochemical complex on the Island."

At the public hearing of July 31, 1964, the Commonwealth Government testified as to the economic difficulties facing Puerto Rico, pointing out that per capita income on the island for 1962 was 50 percent lower than that of the lowest state in the Union. Phillips testified in detail as to its proposed petrochemical project, pointing to a potential investment by the Company and by others of over $597,000,000 and to a direct employment potential of over 33,000 jobs. As publicly stated by the Secretary, "The Department also received testimony and evidence from all other parties desiring to comment on the Phillips proposal, or to offer any other proposals, comments, or suggestions relating to all aspects of the Mandatory Oil Import Program as it affected Puerto Rico." At the hearing of July 31, 1964, statements were received for the record and testimony presented by the following: Sun Oil Company, Embassy of Venezuela, Texas Independent Producers and Royalty Owners, Texas City Refinery Co., Standard Oil Company of New Jersey, Tropical Gas Company, Inc., The Atlantic Refining Company, Hess Oil and Chemical Corporation, Commonwealth Oil Refining Company, Inc., Compania Petrolera de Puerto Rico, Inc., Sinclair Refining Company, Texaco, Inc., Gulf Oil Corporation, Globe Oil Company Virgin Islands, Inc., Standard Oil Company of Indiana, Socony Mobil Oil Company, Inc., Independent Petroleum Association of America, and Tidewater Oil Company. The record was held open thereafter for 26 days for additional testimony.

In addition to an exhaustive analysis of the record by officials of the Oil Import Administration and the Bureau of Mines, field investigations were made in Puerto Rico with respect to the proposed project by the Solicitor of the Department of the Interior and the Oil Import Administrator at the request of the Secretary of the Interior.

2—Mr. Mitchell's letter states:

> "It is hard to believe Secretary Udall's reasoning that the private negotiations with Phillips in Puerto Rico was done in behalf of Puerto Rico's economy."

Public testimony has revealed that the Puerto Rican petrochemical project will provide a very substantial boost for the Puerto Rican economy. We believe that those officials entrusted with economic development in Puerto Rico are better witnesses than Mr. Mitchell to the beneficial effects of the project. In this respect Mr. Rafael Durand, Administrator of the Economic Development Administration of the Commonwealth of Puerto Rico, stated the following at the hearing of July 31, 1964, with respect to the Phillips proposal.

> "It is not overstatement to assert that the Commonwealth of Puerto Rico views the proposal before you as crucial for the future economic well-being of our island. It is the result of three years of effort and negotiation in which our government sought to interest elements of the petroleum and chemical industries in establishing in

> Puerto Rico the core of a Petro-chemical complex which we deem urgently necessary to meet and conquer the economic factors converging to threaten and challenge the industrial development and social well-being of our island."

An examination of the agreement which has been negotiated between the Commonwealth Government and Phillips shows the commitments that the Department and the Commonwealth have required Phillips to undertake in order that the petrochemical project would provide maximum benefits to the Island's economy. Some of the major commitments are as follows:

(1)—A Phillips subsidiary, 75% owned by Phillips Petroleum Company and 25% owned by the Puerto Rico Industrial Development Company, will invest a minimum of $45 million and will make available working capital necessary for the purpose of providing and operating a core chemical facility capable of processing 50,000 barrels a day of unfinished oils.

(2)—The Phillips subsidiary will reinvest in satellite or other facilities acceptable to the Commonwealth Government a sum equal to all profits of the core facility for a period of ten years.

(3)—The Phillips subsidiary will offer to Puerto Rican manufacturers the right of first refusal at competitive prices for products yielded from the core chemical facility which would otherwise be sold outside of Puerto Rico.

(4)—Phillips Petroleum Company will invest the additional sum of $5 million a year in satellite plants each year after the tenth operating year until the total investment in satellite plants of $55 million referred to below has been reached in the event that at the end of ten years from date of start up and operation of the core facility sufficient profits have not been generated to provide investment in satellites of $55 million.

(5)—Phillips Petroleum Company will at the end of six months from the date of grant of allocation accelerate investment by an immediate commitment to invest in satellite plants so as to bring the total so committed to a minimum of $10 million in the event that contractual commitments have not been received for the establishment of satellite plants representing an investment of at least $10 million at the end of such period.

(6)—Phillips Petroleum Company will at the end of 21 months from the date of grant of allocation accelerate investment by an immediate commitment to invest in satellite plants so as to bring the total so committed to a minimum of $30 million in the event that contractual commitments have not been received for the establishment of satellite plants representing an investment of at least $30 million at the end of such period.

(7)—Phillips Petroleum Company will offer the Puerto Rico Industrial Development Company equity participation with it in any proposed satellite plant up to the amount of 25% of the interest available to Phillips.

3—Mr. Mitchell's letter states:

> "The Phillips handout will unquestionably open the flood gates of uncontrollable imports and this present move by the Interior Department will not only cripple the import program, but will ensure added hardship on the domestic oil industry."

The steps taken and the considerations weighed by the Department in connection with the Phillips proposal serve as evidence that the mandatory oil import program will not be crippled nor will the flood gates of uncontrollable imports be opened. Following endorsement of its proposal by the Commonwealth Government, Phillips presented its application to the Department in January, 1964. A feasibility study of the proposal, including manpower potential, was undertaken by the Bureau of Mines and, following representations to the Department by the Commonwealth Government concerning the unfavorable economic outlook for Puerto Rico, the Department announced public hearings to be held July 31, 1964, with respect to imports into Puerto Rico. As a result of the analysis of the hearing record by the Department together with the findings of the Solicitor of the Department and the Oil Import Administrator, a determination was made that establishment of a petrochemical complex in Puerto Rico as envisaged in the Phillips proposal would go far toward accomplishing the economic objectives of the Commonwealth and should be authorized if a plan consistent with the objectives of the Mandatory Oil Import Program could be developed.

As a result of intensive analysis by the Department in conjunction with the Commonwealth Government, the Department set forth certain terms and conditions for accomplishment of the economic objectives of the Commonwealth in a manner consistent with the basic objectives of the Mandatory Oil Import Control Program. The Department announced that if the Commonwealth would enter into a binding agreement with Phillips incorporating to the Department's satisfaction such terms and conditions the Department would take the necessary steps to implement the Puerto Rican petrochemical project.

The Secretary of the Interior has publicly stated that should the Puerto Rico petrochemical project result in increased shipments of gasoline or other petroleum products to the United States, "it would be our intention to seek adjustments to the program to ensure that net receipts in the U.S. of petroleum and petroleum products from foreign areas would not be increased as a result of such shipments."

It is reasonable to expect that the Department will adhere to these same objectives and take similar steps in treating recent applications by others for petrochemical projects in Puerto Rico.

4—Mr. Mitchell's letter states:

> "It should be recalled that in 1961, the same company negotiated helium contracts with the same Interior officials, without benefit of competitive bids. The General Accounting Office later charged that this deal amounted to a $155 million government giveaway."

As to Mitchell's statement "without benefit of competitive bids," the record is abundantly clear that this is untrue. Interior officials furnished "Guidelines and Information for Prospective Participants in the Helium Conservations Program," dated October 25, 1960, to a large number of companies including Phillips, and requested proposals for helium recovery projects. Fourteen companies responded to this request. By press release of August 21, 1961, Interior officials again requested proposals from prospective helium suppliers and Phillips responded to this request with another proposal. Extended negotiations by Interior officials with concerns which had submitted proposals resulted in the execution of four contracts. The Phillips helium contract was the last of the four contracts to be executed and rather closely followed the pattern of the three previously executed agreements except as to price. The price to be paid for helium as set forth in the Phillips contract is lower than the price in any of the other three contracts.

Talk of a "$155 million government giveaway" in connection with the government's helium program has been strongly protested by Interior officials and the four contracting companies as untrue. Moreover, the statement, erroneous as it was, referred to the entire purchases made under all four contracts and not just to those purchases from Phillips as implied by Mr. Mitchell.

The above, Mr. Attorney General, are matters of public record which refute the assertions and innuendoes which bear upon Phillips Petroleum Company in Mr. Mitchell's letter to you.

Sincerely,

Stanley Learned

July 22, 1965

TO: Mr. Stanley Learned
President
Phillips Petroleum Company

I am enclosing a copy of a second letter I have written to the Honorable Nicholas deB. Katzenbach, restating my position in the matter of the Phillips-Puerto Rican decision.

It has not been my intention to single out Phillips Petroleum Company, or any other company, for condemnation, but rather to point out what I feel are the extremely dangerous consequences of the issue in which your company has become involved.

As an independent, I feel the responsibility to speak out against any decision that will have such a damaging effect on the independent domestic producing segment of our industry.

Yours very truly,

Johnny Mitchell

July 22, 1965

TO: Honorable Nicholas deB. Katzenbach
Attorney General of the United States
Department of Justice
Washington, D.C.

Dear Mr. Attorney General:

Please permit me to comment on the letter to you of July 15, written by Mr. Stanley Learned, president of Phillips Petroleum Company, in which he challenges certain statements I made in my letter to you of June 9.

For the record, my letter of June 9, 1965, stands as is, and I here repeat my suggestion that you undertake an investigation of the special arrangement by which Phillips may import some 50,000 barrels daily of products via Puerto Rico. Such investigation seems all the more needed now to reveal whether my statement or Mr. Learned's more accurately reflects the situation. I welcome his taking up the challenge, and by implication, agreeing that the matter is of such gravity as to require your personal attention.

Allow me to deal briefly with those points in my letter which Mr. Learned chose to dispute:

As to his point number 1, Mr. Learned asserts there is no special arrangement by which these imported products will evade the normal excise duty, that the agreement is a matter of public record, and that Phillips was not accorded preferential treatment.

Gasoline imports into the United States are supposed to pay 52¢ per barrel duty. But by bringing this oil from Venezuela first into Puerto Rico, then converting it into gasoline, this duty is evaded. Naturally, no reference was made to this evasion in the published documents, but the fact is that these imports do evade the existing excise tax which Congress intended such product imports to pay. Had this oil been produced in Puerto Rico, escaping the duty would have been one thing; but employing this circuitous manner to evade it is another matter entirely.

While the negotiations were widely publicized, and other companies knew about them and had an opportunity no doubt to negotiate for a similar privilege, the award was arbitrarily handed to Phillips by the Commonwealth of Puerto Rico and the Interior officials. This remains preferential treatment in my books, and represents bureaucracy at its worst to me.

As to point number 2, Mr. Learned takes offense at my question as to whether this award actually resulted from concern with Puerto Rico's economy.

The interest of Puerto Rican officials in this deal may well be genuine, and it might be a fine thing for the Puerto Rican economy. That's beside the point. My point is that arrangements of this kind undermine the imports program and are a dangerous and improper means of helping the Commonwealth of Puerto Rico. I certainly have no objections if Phillips undertakes to build any kind of facility they desire anywhere in the world, and I hope they build two petrochemical plants in Puerto Rico. But I don't want to have it partially subsidized by special privileges to import products here, primarily at the expense of the domestic independent producers. If this sort of thing can be excused on grounds that it helps Puerto Rico, then why not allow a similar arrangement for a petrochemical plant in Appalachia, or Oklahoma or Texas.

My point was simply that the real motivation for this arrangement is to assure an economic windfall for Phillips Petroleum Company, not Puerto Rico, and I do not believe that should occur at the expense of a defense-vital home industry.

As to point number 3, Mr. Learned cites Interior Department assurances that the move will not be allowed to cripple the imports program.

I say to you that any such assurances on the part of the Secretary and his assistant serve only to show how unrealistic and naive they are in approaching this whole matter. Unless Phillips alone is to be allowed this special privilege, there simply cannot be a corresponding cutback of other product imports to keep total products from rising. There just aren't enough product imports currently to cut back to make room for the new product imports that will result from this loophole. It's plain silly for Interior people to keep telling Congressmen and the public that they can cut new loopholes in the program and then offset them some way or other without crippling the program.

As to point number 4, Mr. Learned insists there was some element of competition in the award of helium contracts to Phillips and three other companies by the present Interior officials.

I say that any element of competition was a meaningless one, by the very nature of the negotiations. The Secretary himself later admitted he "made a mistake." The General Accounting Office claimed in fact it was a $155 million mistake, and only recently upped this amount to more than $170 million.

Mr. Attorney General, my purpose is not to condemn any particular company or government official. It happens that Phillips Petroleum Company is involved in what I believe is one of the rawest deals in our generation. I think it is bad for the Administration, bad for the industry, and bad for the country. I am convinced that it will have far reaching effects and that some alternative arrangement should be instituted before it's too late.

I repeat that the Justice Department should show a little concern over this smelly affair. If it does not, I predict that it will not be long before appropriate committee of Congress and the general press begin to show an interest, and that history will not look kindly upon this kind of carrying on at the nation's expense.

In fairness to Mr. Udall, I would like to raise a point for your consideration. There are reports that this arrangement resulted from a political power play by two men who, while technically outside the government, walk the forbidden corridors of Washington like kings, where Cabinet officials, Senators and Congressmen fear to tread. You might consider the possibility that this could explain why Mr. Udall would agree to something which involves so much that he apparently knows very little about. If this is a maneuver by these two potent gentlemen, don't you believe the Administration should get busy and stop it before the President gets blamed personally? Perhaps Mr. Udall deserves to be cleared of personal blame before he goes down in history like the Secretary of Interior under President Harding. This can best be averted if the Administration itself, acting through you as Attorney General, determines the facts and makes them known to all concerned.

Now that officials of Phillips Petroleum Company have evidenced their own concern over your possible attention to this matter, I trust that you will agree with me that it cannot longer be ignored.

Respectfully yours,

Johnny Mitchell

August 27, 1965

TO: Mr. Johnny Mitchell

FROM: The Honorable Russell B. Long
United States Senate
Washington, D.C.

Dear Johnny:

Thank you for your recent letter enclosing copy of certain correspondence concerning the Phillips arrangement to import oil via Puerto Rico.

As you may know, I was very much displeased with this action and I so informed the Interior Department.

You made some good points in your letter to the Attorney General and I regret that his reply was so unsatisfactory.

With the assurance that I will continue to do what I can to bring about a reduction in oil imports, and with every good wish, I am

Sincerely,

Russell

September 9, 1965

TO: The Honorable Russell B. Long
United States Senate
Washington, D.C.

Dear Senator Long:

I appreciated hearing from you very much and I want to thank you for your inquiry into the Phillips-Puerto Rican issue. I, too, regret that your effort was fruitless. A number of inquiries into Interior's actions were made by other Congressmen and Senators resulting from my letters to them requesting such inquiries. In each case, the feeble answers they received from the Cabinet offices were shameful.

You are aware that Stanley Learned, President of Phillips Petroleum Company, wrote Attorney General Katzenbach answering my first letter to the Attorney General. He defended Phillips' position in this special handout. I venture to wager that Stanley Learned received a personal reply in one form or another from the Attorney General, whereas the Attorney General chose to overlook my first and second letter. It would be revealing, I'm sure, if you and the Texas Delegation would inquire of Mr. Katzenbach and obtain a copy of his reply to Mr. Learned.

I predict that time will prove that the Phillips-Puerto Rican decision was a top priority matter, forced on the Interior Department. We have the suspicion that this package was approved before the knob of Secretary Udall's door was ever turned. Mr. K.S. Adams is recognized as a very close friend of Abe Fortas and Oscar Chapman, and what obligation the Administration owes Mr. K.S. "Boots" Adams has never been disclosed. At any rate, the result is a hell of a giveaway.

The end result of this act will lead to further deterioration of our domestic industry and the final effect will not be felt until long after Secretary Udall leaves office, and even after our President retires to his ranch in Texas.

Many people seem to assume that our President, whose roots are deeply embedded in Texas, is not going to let anything happen to injure Texas and its industries. Even though he was reared a Texan, he is now our President, with many other obligations to serve all of the states for the welfare of our nation and its millions of citizens. This obligation to serve the nation as a whole necessarily involves matters that seriously affect Texas and its industries. A decision that affects Texas should be the business of our Congressmen and Senators from Texas. The weakness in the Texas Delegation is that most of our Congressmen and Senators are not inclined to differ with the President, a Texan, on his proposals and forget that they are supposed to defend the best interests of the State of Texas.

What we need, Senator, is men of courage that will stand up and be counted when our state and our industry is hurting, such as Governor Connally has done, even in opposition to policies proposed by the President. Each of these public officials has his duties and responsibilities, yet they can differ on policies and still have a very high regard for each other. This, in my opinion, is statesmanship. Give me a few more John Connally's and a few more Russell Long's in Washington, and a situation like the Phillips-Puerto Rican problem could never have happened. Our rights of survival are slowly being smothered by inter-Cabinet agencies, private giveaways, and shortsighted administrators of international policies.

I want to thank you again, Senator, in behalf of the independents, for your many efforts to stem the growing tide that is drowning the domestic oil industry. I am taking the liberty of sending copies of this letter. I felt sure you wouldn't mind since we are both working for the same cause . . . . the good of the domestic oil industry and the nation.

I feel certain that the approval of the Phillips-Puerto Rican issue is imminent. Silence always seems to be the prelude to such acts in Washington. I predict that the announcement of the approval of the Phillips-Puerto Rican contract will come just as soon as Congress is adjourned.

When the contract is approved, the plight of the independent will be reminiscent of Joseph, Chief of Nez Perce, and his tribe upon surrendering to General Howard in 1877. Chief Joseph said,

> "I am tired of fighting . . . . my people ask me for food and I have none to give. It is cold and we have no blankets, no wood. My people are starving to death . . . Hear me, my people. I have fought, but from where the sun now stands, Joseph will fight no more, forever."

And Lt. Wood, aide-de-camp to General Howard, who took down Joseph's words, stated:

> "I think, that in his long career, Joseph cannot accuse the Government of the United States of a single act of justice . . . . "

To paraphrase Lt. Wood's words, the independent domestic oil industry today cannot accuse the Government of the United States of a single act of Justice.

Sincerely,

Johnny Mitchell

September 13, 1965

TO: Mr. Johnny Mitchell

FROM: Mr. Jack Valenti
Special Assistant to the President
The White House
Washington, D.C.

Personal and Confidential

Dear Johnny:

I have the copy you sent me of your letter to Senator Russell Long.

You and I have been friends a long, long time. My affection and respect for you remains undiminished. That's why I was surprised at the rather bizarre statement you made—"what obligation the Administration owes K.S. Boots Adams has never been disclosed." There is implicit in that statement a kind of mysterious alignment between Mr. Adams and the government, in which Mr. Adams seems to be a favored person. This is ridiculous as well as totally untrue. Of course, you know this.

I am not aware of the facts concerning the case. But I do know whatever the facts, they will bear plainly no imprint of a "deal." I don't know Mr. Adams. He has never been in the White House for a meeting to my knowledge. He has been accorded, again to my knowledge, no more or no less than any other citizen. What you imply, Johnny, is something that I know you truly cannot believe.

Sincerely,

Jack Valenti

September 23, 1965

TO: Mr. Jack Valenti
Special Assistant to the President
The White House
Washington, D.C.

Dear Jack:

I received your letter of September 13, 1965, and I appreciate your concern over my letter to Senator Russell Long.

Jack, you and I have been friends for a long time. You've come a long way and I am proud of you. As for myself, I have fought the oil industry wars for many years. At times I feel I could hang up my hat, take off my coat, and retire. Instead of following the easy path, I feel a deep obligation to serve this industry. What I give back to the industry is in appreciation of what this industry has afforded me.

I don't need to tell you that anything I may say in defense of this industry, or the independent segment of it, would never be intended as a personal affront to our great President. I've been for him all the way, and I'm not about to change. Nor do I expect him to jump in and do things for us that others couldn't or wouldn't do. But I do reserve the right to express my views when mistakes are being made, like the Puerto Rican-Phillips affair, by any Administration.

You and I know that I'm no diplomat, and I don't try to be. I call 'em just as I see 'em. I strongly feel that's what the President needs most from his friends. The worst disservice we could do him would be to say that whatever his Cabinet or his Administration does is right, or shouldn't be criticized because he is a Texan or because he understands our industry or because he's a friend.

The President, even as effective as he is, can make mistakes, and he will get blamed when some in his Administration make mistakes. I think a mistake is being made in the Phillips deal. It's bad, it's unfair.

Jack, I'm never going to stop supporting President Johnson personally. But I'm also never going to stop criticizing things I think are wrong. It's my nature and I couldn't change it if I wanted to. I honestly believe I'm working for the good of my industry. Maybe I'm wrong, but I have to do what I think is right and best. Consider the possibility that I may be right . . . . then the Phillips-Puerto Rican deal may look a whole lot different when you do.

Kindest personal regards.

Sincerely,

Johnny Mitchell

September 30, 1965

TO: Mr. Johnny Mitchell

FROM: The Honorable Olin E. Teague
The United States House of Representatives
Washington, D.C.

Dear Johnny:

I suspect that I would be better off if I kept my mouth shut and did not send you this letter. However, with the hope that it might be helpful, I am sending it.

Yesterday, listening to a number of Members from Texas discussing the depletion allowance, your name came up; and the remark was made that someone, somewhere may be able to tell Johnny Mitchell something, but they are not going to be able to tell him much because he already knows it all. The people talking did not know that I was acquainted with you. The people talking were people who are very interested in maintaining the depletion allowance; the people talking were, on the whole, people who have opposed the Administration on the majority of the issues in the Congress so they were not among the group whom you say is "the weakness in the Texas Delegation" in that "most of our Congressmen and Senators are not inclined to differ with the President, a Texan."

The people I am talking about, Johnny, know the problems of the depletion allowance situation in the Congress from "A" to "Z." It has been more of a problem I think than you have ever realized or recognized because if you people in your business only knew, you would be forever grateful and patting the backs of the Texas Delegation for having saved the depletion allowance back through the years.

I know through personal knowledge that Sam Rayburn did more to save the depletion allowance than any other living human, when the very industry involved except for a few people were kicking the hell out of him every chance they had—and there were many chances.

Thus, the point of this letter is about the Members of the Texas Delegation who are very sympathetic toward the very things in which you are interested and believe that you and those you represent are going to make a tremendous mistake as far as the depletion allowance is concerned. So, Johnny, what I want to suggest is that you and four or five of the top people in your industry come to Washington and I will arrange a luncheon, dinner, cocktail party, or a get together of some kind where this problem could be discussed with the Members of the Texas Delegation who believe the same as you do but who know that it is not nearly as simple as you seem to think. For your information, we in Texas have a good Delegation in Congress whom I think are very representative of our state.

I doubt, Johnny, from reading parts of your letter to Russell Long, you ever dreamed how sensitive the subject of depletion is. I doubt you dreamed of how much the Texas Delegation has done to preserve it and how very, very easy it would be to lose it. If you are interested, let me know. As I am sure you know, I am not as knowledgeable of the oil industry as I would like to be; but I am knowledgeable enough to know something about the depletion question in the Congress of the United States.

Sincerely,

Olin E. Teague

October 15, 1965

TO: Mr. Jake Hamon
P.O. Box 663
Dallas, Texas 75221

Dear Jack:

I have just returned to Houston and become aware of your letter of resignation to TIPRO. I am not as concerned about your resignation as I am about your feelings on independents expressing their opinions, either emotionally or with great deliberation, on foreign tax credits, foreign depletion, or any other industry problem that effects the independent.

I can assure you that I have as great a love for this industry and hopes for industry unity as anyone else. You have the same feeling for you have mediated the cause of the independents and have been the buffer between the independents and majors for a great many years. However, your independent ties are coupled with a responsibility to the majors. I do not feel as strongly as you in this respect, nor do I feel the same ties and responsibility to the leaders of our government as you. Because of your feelings toward each of these factions, they constantly call on your judgment and abilities to smooth the troubled waters between independents and majors, and it is only natural that you become deeply concerned when issues arise that are certain to rock the boat of unity.

I, too, feel a definite responsibility to the industry. However, my purpose and my obligation is dedicated to the survival of the independent producers.

You must realize that we are being closed in tighter and tighter by monopoly trends of the major oil companies and their selfish motives. The ambition of each major company recently has become so avid that they are even playing cut-throat games with each other. When this happens among the giants, where do you imagine a place can be found for the small independent?

Our good friend, Michel Halbouty, took up your cry. He, too, is a good friend of mine; yet Michel, like yourself, has included an obligation to major company officials as a part of his requisite for success. It takes all breeds of independents, with their various philosophies, to succeed. Each has his right to fight for the principles he believes in.

As for myself, Jake, I have long been a proponent of depletion and intangibles, and I sincerely believe that our depletion allowance is too low for the risks involved. Regardless of the way I feel, there is a special foreign tax privilege allowed the giant international oil companies, both in royalties paid to foreign nations, in their intangibles, and in depletion, contrary to Halbouty's unfounded statements. Their tax credits are handled in numerous ways to their advantage so that the end result is that hardly any taxes are paid by any giant international oil companies operating abroad.

Jake, when you begin to believe that we can't or won't be sacrificed to the wolves by the majors, then you are not facing the situation realistically. We cannot live in the past, nor can we sit back and expect the industry to protect us. We must fight for every inch of ground that it takes for our own protection.

In recent months you have argued in behalf of decontrol of residuals without realizing, apparently, the drastic effect it will have on the crude import program. You were concerned with the attitude of the East Coast Senators who could care less about anything other than their own part of the country. Where they would like to decontrol our life's blood, these same senator's made certain there were import duties and restrictions on all mercantile goods.

While you were so strongly advocating decontrol of residuals, did you ever stop to realize you were playing into the hands of Standard of New Jersey who are the prime movers of this decontrol for this is their territory for fuel oil supply?

Standard of New Jersey also controls the greatest unsold gas supply in America, which is located in Texas. Where they should be trying to sell this gas energy to the needy market on the East Coast and the Mid-west, they are purposely trying to divert this gas to the West Coast which is already over-saturated with gas supplies.

Did you realize that the West Coast actually produces some 400,000 barrels of low gravity crude oil per day which could have been without a market because of excessive gas imports into District V had it not been for Standard of California and their research on their Isomax unit? Did you ever realize that Standard of California, Union, Richfield, Signal and other major companies on the West Coast have spent millions of dollars to construct Isomax units and hydrocracking plants which will convert this otherwise useless low gravity fuel oil into gasoline and other products, thus making use of the independents low gravity oil production? While the majors on the West Coast are trying to save the low gravity fuel market for the small independent Standard of Jersey, you and others are advocating an open door policy on fuel oil.

Let's be realistic, Jake. Let Standard become industry-minded and send her gas where it belongs. Forget the feelings of the unfriendly senators from the East Coast, and let's save our own necks for a change.

With regard to the many statements made by Michel Halbouty in his recent letter, only a few of which were constructive and the rest obviously major company concepts, the most important were the remarks on foreign tax credits and foreign depletion. We may have to look to these two sources if we are forced to. If we are forced to, there will be no regret on my part at bringing these issues up and laying them before Congress and the public. Even if the attack on foreign depletion and foreign tax credits did cause an investigation into domestic depletion and we could not defend it, the fault would still lie with the selfish major companies that have caused the breach in industry unity.

My final comment to you is what good is depletion to an independent producer when it is possible that in the future he could have no oil to deplete. And from the way things are doing in this selfish trend of major international oil companies, it is evident that the oil we will have to deplete will grow less and less each year. As far as I am concerned, although I believe in depletion, today it offers very little incentive to the new oilmen when they must find new oil at prohibitively high costs, and sell their oil at a low price, solely because of excessive imports and foreign tax advantages of the major international oil companies.

If these companies cannot see the hardship they are creating for us, then instead of rising to the call of "Don't rock the boat on depletion," it will be my obligation to my fellow producer to join the independents that will rock the boat, and it will be my pleasure to pull the plug that will sink the boat of unity.

Yours very truly,

Johnny Mitchell

December 1, 1965

TO: Mr. John Doe
Washington, D.C.

Dear Mr. Doe:

I believe the time has come for somebody who isn't running for or from anything to talk frankly about the oil industry mess in Washington.

If you're wondering why I'm addressing this letter to you, Mr. Doe, it's because everybody in a position of authority there is either swearing off oil issues or has demonstrated an incapacity to understand what's happening. Let me tell you what's happening to the independent oilman.

Some say the President is playing it really smart by letting out the word that Udall is making all the oil decisions for his Administration. But if he should ever ask me, which isn't too likely, I'd tell him it's a mistake. He's going to get the blame anyway, and unless he does take charge there isn't going to be any credit for anybody to take. I can't help thinking a better way would be for the President to take charge, tell everybody he's in charge, start making oil decisions on the basis of what's best for the country, and then watch the oil-hating politicians run for cover. I think the oil industry has been the whipping boy of the politicians long enough, and that the President owes it to the industry to face them down once and for all. What could the President lose? Plenty, if he's making oil policy on the basis of expediency. Nothing, if he's making oil policy on the basis of what's good for America.

Well, anyway, that's why I'm not writing the President these days. You may ask why I don't write some oil-state Congressmen. I tried that too. Found they're all afraid somebody might get the idea they're criticizing the President if they do anything that hasn't been cleared.

A few months ago, in frustration, I decided to write the Attorney General, on the naive assumption that he'd be interested in hearing about the monopoly trend in this industry. I thought he'd also like facts about a little situation that's going on that could do for some selfish oilmen what Teapot Dome did for old Harry Sinclair . . . . get them jailed.

And you may say, if it's about oil why not write Secretary Udall? He keeps telling everybody he's running things there. You're right, he does. And that's what I want to talk about.

This fellow Udall is a nice enough fellow, really. I read his book about a "Quiet Crisis." Sure enough, he's for doing good and all. Trouble seems to be mostly that he can't keep his mind on oil problems except every six months when he has to divide up import quotas among a lot of grateful companies. And now he's got that changed to once a year.

Not that he isn't a practical man. Take the Phillips deal in Puerto Rico. Claims that's all his idea. Even leaves the impression he twisted the President's arm to get him to sign the special Executive order handing Phillips the juiciest oil plum ever awarded an oil company by a politician. Even made it a ten year deal to be sure. 50,000 barrels daily feedstock guarantee, with special rights to bring gasoline by-products in from Puerto Rico. Had to cancel out his earlier plans to eliminate non-residual products to do it, but he did. Also left the Puerto Rican imports outside the 12.2 percent ratio limit which a lot of folks think covers all imports. Said it was because Puerto Rican economy needed a boost.

There's a lot more to that story, and we haven't heard the last of it yet. Major companies will see to that. They didn't like it and were brash enough to say so in public. One major company official was even uncouth enough to circulate the names of Administration intimates who handle such deals for Phillips and Puerto Rico.

This import situation is pretty fouled up. Take the latest action on residual fuel oil. The problem all along has been an allocation system which gives quotas to people who only want to sell them to somebody else. Seems folks would rather make a windfall out of a government quota handout than work for a living. But Mr. Udall and the boys never got around to changing the system. Instead, not long ago he threw up his hands and came out for decontrol. This must have been a lapse on his part, because decontrolling things isn't a specialty of this Administration. So now he has gone back to raising quotas again. He did it again the other day when stocks were at an all-time high.

Crude oil imports handling isn't much better. Idea behind import controls was to keep cheap imports from squeezing domestic producers out of business of drilling for new oil—to keep reserve capacity at home to meet needs in emergencies. So they set up a quota system. Didn't work, but better than nothing. Since Congress adopted the security clause for oil, imports have more than doubled while domestic crude production has increased an average of less than one percent annually.

But that's not the worst of it. Letting company quotas be handed out on the basis of bureaucracy rather than economics tends to get pretty messy. To keep independent refiners—who didn't own any foreign oil or couldn't even use any—from going down the drain under competition from cheap oil in hands of majors, they doled out quotas to them too. Had to jiggle the formula to allow higher ratio to small refiners. But at least they got independent refiners quiet. Some doing so well, in fact, they're threatening the majors' business.

After the Phillips Fiasco, the petrochemical people got wind of a possible quota handout, and began threatening to build plants abroad if they didn't also get a quota subsidy. Clinched their case when they added that handing them this subsidy would increase their exports, and help the balance of payments. Mr. Udall apparently thought this was Great for Society. So he broadened the allocation base again to include petrochemical plants. Only Phillips is happy, though hopeful there will be no Senate-investigation. The rest aren't happy. They claim his handout wasn't near big enough to keep them from building their plants abroad and won't subsidize more export sales after all. So all we got was an increase in imports to appease people who don't aim to be that easily appeased. Meanwhile, Mr. Udall got some temporary mileage out of some loose talk at his press conference about "freezing" something or other, and a few IOCC people who hadn't done their homework thought he had frozen the level of imports so domestic production could get a little growth.

About this time I got a notion that independent producers might get included in the Great Society. Since Mr. Udall is so fond of imports, and quotas seem to be his answer to everybody's troubles, why not include the people Congress intended to protect by import curbs in the first place? Sort of "go along to get along" if I may coin a phrase. Well, I developed a plan to do this and sent it to all the major company brass. Got a lot of interesting answers, all adding up to the same thing: "Johnny, if you don't stop messing around, you're going to screw up our whole imports program." Here I'd been operating on the theory that they's beaten me to that.

If things are that bad, you may be saying, surely there's some prospect for improvement. Not likely. You see, Mr. Udall thinks things are fine. Unfortunately, only the majors are making record earnings.

So what about it, you may be asking. What if the independent producers are going under during peak prosperity for the nation? After all, they're oilmen, and all oilmen are rich anyway. I'll tell you what about it, Mr. Doe. When the independent oilman is hurt, the domestic wells don't get drilled and the oil doesn't get found. Let this go on a few years, like it has, and our reserve productive capacity goes to pot. If any significant portion of foreign oil is cut off to our allies, like in the Suez crisis only a few years ago, we'll be helpless. It's that U.S. reserve productive capacity on which the free world "floated to victory" in recent wars. And it's going fast. So just cause the majors are making lots of money doesn't mean the oil industry is doing all right, Mr. Doe. Wouldn't take much to correct the situation, but sure doesn't look like Mr. Udall is going to bother about it.

We have other troubles, Mr. Doe, and I may tell you about them in my next letter. Meantime, if you run into the President, you might just casually mention that we aren't very happy.

Yours very truly,

Johnny Mitchell

February 9, 1966

TO: Congressman Olin E. Teague
House of Representatives
Washington, D.C.

Dear Tiger:

I want to express my deep appreciation for your letter in September, and for your suggestion of visiting with the Texas Delegation in Washington. You realize, I'm sure, the depth of our long friendship. I may write in my many letters that our Texas Delegation may not be as forceful as I would want them to be, but this has no reflection on the way I feel about you.

The only reason that I make these public expressions about the Texas Delegation is that it gives them the opportunity to defend their position if at any time they wish to use my accusations as a stepping stone to their defense. In doing so, the Administration and the President could only jump on me, which I could care less about.

Remember, Tiger, I am a good Democrat, I have always been one, I have paid my share towards supporting this Administration and, unlike most supporters, I have not asked for a favor in return. So, if I feel that our industry is being by-passed, I feel at liberty to say so, but this doesn't alter my feeling for each of my Congressmen or will it alter my continued support of this Administration and the President.

Sincerely,

Johnny Mitchell

May 1, 1966

TO: The Honorable Stewart L. Udall
Secretary of Interior
Washington, D.C.

Dear Mr. Secretary:

I am addressing this letter to you, man to man . . . from a man who has a great love for the domestic oil industry, to a man who is responsible for the governmental policy affecting the oil industry.

Attached are the essential points of a new oil import plan which, while preserving the equities of all concerned, will enable the Oil Imports Program for the first time to achieve intended objectives.

Under this proposed revision, no company nor source country concerned need be injured, and the glaring inequities of the present arrangement will be lessened without undermining the essential structure of the existing program.

As you are aware, crude oil import quotas today are allocated primarily on the basis of refining-input, and need bear no relation to domestic production whatever. Yet the purpose of import controls is primarily to preserve a healthy domestic producing industry capable of meeting the nation's essential oil requirements in war or peace. The refiner-input theory of distributing quotas to all refiners (including independent or inland refiners which have no foreign oil and would not normally be importers) was that the economic advantage provided by the quota would be passed on to the domestic producer in the form of crude prices adequate to maintain exploration and drilling. It hasn't worked. Our government acknowledges its failure by declaring that wildcatting at home is falling seriously short of minimum safe levels.

Nine years of woefully inadequate U.S. oil exploration and development activity is threatening to catch up with us . . . and in a hurry. A few short months of significant demand increase combined with escalated Viet Nam war needs now reveal for all what knowledgeable domestic oilmen have known for a long time. Our nation's reserve productive capacity has taken a beating and is currently in no shape to meet the terrific demand and security requirements estimated for the coming 15 years. Obviously, a change is needed. I say the change should be made as soon as possible.

To repeat, the change proposed here would accomplish the objective of restoring exploration and drilling at home adequate to maintain our nation's oil self-sufficiency, while doing no damage to importing interests and source countries.

I recommend first that non-residual fuel imports be restricted to a total of 1.5 million barrels daily. Secondly, I propose that 20 percent of this total be set aside to provide quotas to producers who drill for oil in the U.S., with the reservation that companies now importing oil must choose either to keep their current quota rights or participate in the new drilling quota plan, but not both. The more feet drilled either in wildcat or development wells by a producer the more quota he is eligible for. I include development wells not only to make my plan more workable, but also to recognize the fact that development wells are equally important in providing reserve productive capacity. Since most producing states now operate under sensible conservation rules regulating spacing and other drilling requirements, I am assuming these new quota privileges could not be abused.

I further assume that companies which now import and also drill in the U.S. would not find it profitable to give up their current importing rights and participate in drilling quotas. Consequently, virtually all of the 300,000 barrels daily set aside would be preserved for domestic independent producers who need the incentive required to solve our reserve problem.

Last year approximately 178,000,000 feet of hole was dug in the United States, and 35 percent of this was dug by the independents, amounting to 62,500,00 feet. If we use the 300,000 barrels per day of import quotas as an incentive for the independents, it could amount to $108,000,000 per year, judging from prevailing values in quota ticket trading. This, for every foot of hole dug by the independent, would amount to $1.75 per foot, and I feel sure this is a sufficient incentive for us to go back to work.

In considering the amount of imports to be allocated under this plan on a ratio with domestic production, you will note exclusions of (a) residual fuel imports, (b) independent refiner quotas, and (c) overland exempt imports. Why?

Residual fuel imports are deducted from the total, not because I am recommending decontrol of residual fuel imports, but simply as a recognition that these product imports are presently treated separately from other imports. (I personally believe it is possible to limit resid imports to actual needs of reside-to-be used-as-a-fuel, assuring adequate supplies while avoiding over-supply, without the necessity of decontrol.)

Independent refiners will continue to get quotas on the same basis as at present. I do not believe it is necessary or desirable to disturb their arrangement. (Whether their present quotas are enough or too much to enable them to pay the going price for crude in competition with the integrated majors is something that can be considered from time to time. There must be merit to every phase of any import program; assuming their present arrangement stands that test, it should not be changed.)

Overland imports from Canada and Mexico are deducted also. That is not to say that concern need not be expressed with the disproportionate growth of imports from Canada, and attention will be required to this problem. I am saying only that the quotas assigned domestic producers on the basis of their individual production need not reflect Canadian imports in order to provide the relatively small but desperately-needed incentive for exploration and drilling at home.

I will appreciate your taking time to review this plan, and I will look forward to your comments.

Yours very truly,

Johnny Mitchell

May 29, 1966

TO: Mr. Carl Reistle, Chairman
Humble Oil and Refining Company
Houston, Texas

Dear Carl:

I find it difficult to accept the news of your retirement. As this letter reaches you, you will be clearing your desk drawers, carefully picking up each momento of the past from your desk top, and as you carefully take each plaque of honor off from your plaque filled wall, the cherished memories of the past will flash by.

Retiring from a long busy career as Chairman of America's largest producing oil company has to be one of mixed emotions. The rewards and accolades as an engineer and a leader has earned you a spot of recognition even in retirement. Retirement is usually the hardest to take but for you it will be a path well travelled by many friends.

Carl, as you walk down that hall for the last time, I want you to know how the independent segment of this industry feels about you. Your company, for the past few years, has been the umbrella watching over the small independents. Our nation, particularly Texas has been plagued by excessive imports of oil.

Texas, since the beginning of market demand proration in 1930, enjoyed a 30 day producing schedule. Excessive imports reduced the need of Texas crude to eight producing days each month. Independents have gone broke by the hundreds, falling out like flies.

It became a habit watching your Humble in action at the monthly statewide hearings. When it became obvious that company's after company's nominations were below an eight day schedule, you would increase your nominations to equal an eight day demand. I believe many thought that Humble always needed the extra crude, but most of us knew better.

Things got progressively worse in the early 60s. Imports kept increasing. Major purchasers kept reducing their nominations and finally pipeline connections were cut off in several areas in Texas. This found small independents with their tanks full of oil and no buyer. You would always ask, after the hearing, where was the distress oil and how many thousand barrels were involved. Your company bought it all, and it wasn't even in your purchasing territory. You saved a great number of small independents and they never knew you had a hand in saving them.

Carl, back in the 60s there was a lot of territory separating a small independent and the Chairman of Humble. Why you took an interest in me, I still don't understand. I was always a problem to all majors with my speeches, usually against government and major oil companies. I won't forget my caustic speeches at Commission hearings, and my usual remarks about major oil companies at the National Petroleum Council meeting in Washington. I was always concerned about your reaction to my outbursts and believed that you would eventually give up on me. You never did Carl, at the end of each meeting in Austin and in Washington you would have a folded note handed to me quietly in the crowd. The note would usually say . . . "You are the dumbest S.O.B. I've ever known. You'll never learn to keep your mouth shut. My plane leaves at 3:30 p.m. Be there." Then on board, in front of all, you would preach to me like a father to a spoiled son. You always ended up saying . . . "Who in the hell do you think you are? . . . You can't control Congress . . . You can't change the thinking of the Executive Branch . . . And you can't influence the majors . . . Why don't you give up?" I never believed you really wanted me to stop, or I would have lost you long ago.

Carl, I'll always keep you locked up in my special memories. I hope that all the future years ahead are the best years of your life and Mattie's.

My deepest respect always,

Johnny Mitchell

July 12, 1966

TO: Honorable Thomas H. Kuchel
Senate Office Building
Washington, D.C.

Dear Senator Kuchel:

I read with interest your demands for an explanation from Secretary of the Interior, Stewart Udall, for the movement of 10,000 barrels of gasoline, duty-free, from Puerto Rico to the West Coast.

Those of us in the oil industry, who have long waged a lonely and losing fight against detrimental governmental policies, welcome your aid. You are right when you say this latest importation of 10,000 barrels daily of gasoline jeopardized California's competitive marketing economy.

If California refiners are forced to drop their marketing prices to compete with tariff-free Puerto Rican gasoline, great damage to the California domestic industry will result. In fact, so damaging is this gasoline importation program, that I can only believe it is a typical bureaucratic stunt that will favor a few who have paved their way in Washington with favors and promises.

Assistant Interior Secretary J. Cordell Moore's glib denial of your charges is just another indication that this Administration is determined to keep, regardless of cost to the domestic industry, a deal it made to give extreme preferential treatment to companies. . . . . .one in particular. . . .which import their foreign oil here through Puerto Rico. Mr. Moore will give you no satisfaction, because he cannot. If you simply want to make an impression on California refiners, you can carry on your correspondence with Interior officials who really don't have anything to say about policy at that level. But if you are serious in wanting to do something about a situation that is already bad and due to get much, much worse, then I will tell you how.

As a Republican Senator, you must surely know that you aren't about to reverse a Democratic Administration handout by simply carrying on a public correspondence with department officials. But if you care to use your considerable influence as a member of Congress, then there would be some chance of success.

In case you are not aware of it, the future problem that will make your present 10,000 barrel import into California look like a drop in the bucket relates to a deal made with another Republican. When the results of that arrangement are realized soon, imports of gasoline from Puerto Rico will be many times the present level. . . .enough, in fact, to hurt the whole price structure of the board of Phillips Petroleum Company, who, with some wellplaced insiders, walked off with the biggest prize ever handed any company or individual, including the one that led to the Teapot Dome scandal.

A number of us asked serious and probing questions before this deal was signed by the President. Almost all the other major integrated companies declared they would want a similar plum, if Puerto Rico was going to be made the playhouse for Phillips. As you know, Puerto

Rico doesn't produce any oil. It's just a convenient port of entry for foreign oil, and when refined there, escapes the 55 cent per barrel gasoline tariff that would apply if imported direct from other countries.

Now, Senator, let me tell you how to make your opposition to the Puerto Rican import deal effective, assuming you are serious. You have shown courage and determination, and this is something frequently lacking in the present political climate. I, a Democrat, credit you, a lone Republican, with courage and statesmanship to do what you have already started. But where are you?

You have demanded an explanation from Secretary Udall, and got doubletalk from an assistant secretary. You can tell the California refiners you tried, and quit. . . .or you can do something which can get the job done, and thereby help the whole domestic industry. Here's how:

> As a member of Congress, you can lead a movement for legislation. That may sound like a big job, but right at the moment there is a way to make it fairly easy. Just the other day the Senate Finance Committee voted to limit watch movements from U.S. territories. . . the Virgin Islands, Guam, and American Samoa. Their problem is a direct parallel to the Puerto Rico oil import problem. So why don't you press for an amendment to the watch bill which would limit oil movements into this country via Puerto Rico. I don't mean anything harsh, just set some reasonable percentage limits so that the Phillips deal can't completely wreck the home market. The bill on watch imports via U.S. territories would cut back movements by as much as 75% from the Virgin Islands. If you move now, it won't be necessary to offer anything so drastic for oil. Just see that Puerto Rico doesn't suddenly become a hog by limiting its share of the U.S. gasoline market to a reasonable percentage.

If you really want to make an impression on the Administration, offer a reasonable amendment along those lines. This would dramatize the ridiculous nature of the Phillips deal, and its future consequences.

I think I can guarantee you that if you will get this thing before Congress, and get hearings on it in Congress, the Phillips deal and its damaging consequences will fall through fast. Hearings on it last year at Interior found all the independents and nearly all the majors strongly opposed, but the Administration kept the Phillips deal anyway. But hearings in Congress will be something else. Some embarrassing questions will have to be answered. I don't think Mr. Adams will want that kind of attention.

This is the only way to stop this thing. It won't be done by issuing press statements from your office telling Californians that a lone Republican is battling with the Administration on oil imports. By going the legislative route, and getting a public hearing, you'll find yourself backed by all domestic oil interests and almost all the major international companies as well. Opposing you will be only a few with special interests in the Phillips deal, and they won't be too anxious to testify. With odds like that, even a Republican might score.

Respectfully yours,

Johnny Mitchell

October 12, 1966

TO: The Honorable Craig Hosmer
House Office Building
Washington, D.C. 20515

Dear Congressman Hosmer:

I have reviewed the speech you made recently in Long Beach, California with great interest, and commend you on your knowledge of the problems of the oil and gas industry in California. Your criticism of me has merit, for I, like many others in the oil industry, expect our Congressmen to know all the answers without presenting the facts. We know our problems, but it is our shortcoming in not taking the proper steps to explain them to our Congressmen. We appreciate your taking time to find out the facts for yourself.

You may not be aware of it, but since your speech you have joined the ranks of other distinguished industry statesmen. To qualify for this privilege, you have to be criticized by California's most active industry association, Oil Producers Agency of California. I read this agency's comments on your speech in the Oil Daily the other day, and you are now elevated to the same statesmanship status in our industry as Jack Rathbone, past chairman of the Board of Standard of New Jersey. He, among others, has also been criticized by this same agency when anything is said to disturb the status quo situation of the California producing and marketing industry.

In late 1964, Mr. Rathbone made a statesman-like analysis of what has happened to the West Coast oil industry by its insistence on being isolated. At that time, the industry was startled when Stark Fox, Vice President of Oil Producers Agency of California, issued a press release criticizing Mr. Rathbone's appeal to the industry.

I don't know whether you realize it or not, but from the earliest days of the voluntary and the mandatory import program, Oil Producers Agency of California has been preaching the doctrine of isolating the West Coast. The fruits of this policy include a serious depression for independents in California. Only a handful of independent producers are still able to even search for oil out here. The oil we do produce here, we have to sell at distress prices. . .on terms that suit the importing companies, who are also the purchaser of the independents' oil here.

In his published memorandum dated September 14, 1964, Fox claimed that since most of the increase in imports comes from Canada, there's nothing to worry about. He says the Federal government considers Canadian oil as domestic oil insofar as national security and imports policy is concerned.

I say that no knowledgeable government official believes that. You don't believe it, I surely don't believe it.

In the first place, Canada itself is a big importer of oil. They import about as much cheap foreign oil on their east coast as they are forced to dump on us from their surplus-producing western provinces. As a matter of fact, the same international companies who import here have it rigged to guarantee Canada that anytime they increase their imports of Venezuelan or Middle Eastern oil into Canada, they'll dump an equivalent amount of Canadian oil on the U.S. market. . . most of it into District V.

It is true that Canada was given an exemption from import quota controls on grounds that this imported oil is safer because it moves by pipelines from Canada. But in an emergency, Canada would need all her own oil and more, especially if her imports were cut off or cut down.

Even if Canadian oil were as reliable as domestic oil, is that a sound reason to keep letting Canadian imports grow so much faster than domestic production? Mr. Fox took the ridiculous position, in effect, that Canadian oil is more secure than U.S. oil.

If Mr. Rathbone was "naive and irresponsible," as these California "independent" spokesmen claimed, then so are a lot of other people, yourself included. In fact the whole idea of an import program, judged by their standards (of what's good economics for importing interests) is equally "naive and irresponsible." And the ranks of the "naive and irresponsible" seem to be growing.

You were criticized in The Oil Daily for your statement regarding my remarks about the independents being at the mercy of the international integrated companies who control almost all of California's production, all of the imports, and most of the refining and marketing in this state.

Oil Daily's Bill Gregg made this comment, "California independent producers, with few if any exceptions, are extremely proud of the fact that the industry has operated successfully under this state's voluntary conservation program (which gears production to the MER's of individual flush and semi-flush pools to eliminate underground waste) rather than under government-set monthly production limitations such as prevail in Texas and other states."

Now he knows better, and the OPAC knows better, but neither has the courage to admit the truth nor do they dare admit the truth. For we independents have been at the mercy of these monopoly minded major international oil companies for too many years. They have triggered our crude prices to suit their demands and have abolished the Laws of Reason in this industry. The old adage that Supply and Demand govern crude prices is no longer true in California.

They both remember the trying days of the California independents when their wells were shut in looking for a market, while the offset wells belonging to the major international importers were producing at this so-called MER, which only fits the purchasers' purpose.

Contrary to OPAC's statement, other states where regulations exist, every barrel of oil produced by the smallest independent finds a market, and at the market price. Every independent surely remembers when California oil owned by independents went begging for a market, and if a market were found, their oil usually sold at distressed market prices.

OPAC may be proud of this state's voluntary conservation program, which you say is inadequate, but I can assure you many of the abused independents in California will agree that you are right.

I wish to compliment you on many points in your speech, among these, you recommend that possibly the independents should organize into a truly independent association, which we do not have in California at the present time. Such an association is a possibility and, I think, a necessity. Whether there are enough true independents left, remains to be seen.

Yours very truly,

Johnny Mitchell

February 13, 1967

TO: Mr. R.C. McCurdy, President
Shell Oil Company
50 West 50th Street
New York 20, New York

Dear Mr. McCurdy:

It isn't often that a small matter such as this be referred to you. However, what is routine in the daily workload of an international company is important enough to me as a small independent to write to you.

Just last week the National Petroleum Council came out with the finding that the independent producer, once the historic oil finder and wildcatter, is no longer a key figure in the domestic oil industry. There are reasons for the independents' problems in this industry. The Administration and the majors share the blame for premeditated handouts, unfair marketing practices and excessive imports. These are the factors that contribute so heavily to the low price of a barrel of crude.

What also hurts the independent is the uncooperative attitude of major companies. This has done as much to reduce wildcatting and exploration development as the price of crude. Major companies at one time were receptive to the independents' efforts to explore, were always helpful in either dry hole contributions or acreage contributions. They also considered that the cost of the venture should be proportionately shared. These policies have changed, for now major companies not only decline to cooperate, but hardly ever share in ventures of mutual interest.

The blame must lie in leadership. As a typical example of what I am talking about, I refer you to the enclosed copies of correspondence between Jade Oil & Gas Co. and your Shell Oil Company district office in California. Jade owns a 200 acre lease in the prolific McKittrick area in California and we plan to drill a costly well to 10,500 feet. The information to be derived from this well is as beneficial to Shell as it is to Jade. Whether we are wise to drill such a well is what makes the independent producer what he has always been. . . .the nation's "wildcatter." It takes a lot of guts to even attempt such a venture.

I believed that when I made my drilling obligations, Shell would either make a dry hole contribution or contribute some acreage. My relationship with Shell didn't begin yesterday, but dates back some 25 years.

The first reaction of your district man to our request is typical of all employees of major companies in the industry today. "Why should we help or contribute because if we help, and make a dry hole, it could be a mark against my district." The letter from Shell was very cordial, stating that the subsurface geology did not merit any support. To me this could only mean that the acreage Shell owns had to be a little value. With this admission, I then asked for an acreage contribution. The second reply from Shell implied they did not want to contribute acreage either, but invited us to come back and talk with them if we found production. One can imagine the reception we would get from Shell if we were successful in this wildcat venture.

It doesn't matter, for we will go it alone. We independents have become accustomed to this same reaction from every major company. A wildcat venture such as we are undertaking at McKittrick is the typical dream and hope of the independent. Once these dreams and hopes were also shared by major companies. Today, even though the prosperity of every major company has exceeded all expectations, the joint ventures we independents once shared with the majors no longer exist.

I am writing this letter to you not so much to make an issue with your company but, as an example, explain to the industry that one of the primary reasons for the independents' decline is that the major companies are not sharing their part of the risk to explore, signifying they are not concerned enough with the nations declining reserves and the nation's security.

Where should the responsibility be placed? Only one place. . .on each executive officer of each major company.

Yours very truly,

Johnny Mitchell

February 27, 1967

TO: California Delegation
United States Congress
Washington, D.C.

Gentlemen:

We could write each of our California Delegates individually, but the California independent oil producers' future is so critical that all of us should know about a serious new problem that we are facing at this time. We are asking you to investigate recent Administration action that could further reduce exploration in California. As it is today, we are not only failing to find essential future oil reserves, but are finding it almost impossible to produce what oil we now own. At present day crude prices in California, there simply is no room for profit.

A few weeks ago some of the major international oil companies raised gasoline prices by one cent per gallon. Unprecedented Administration threats prevented others from doing the same. This nominal raise is long overdue and benefits not only the major refiners, but the independent refiners, brokers and independent marketers. The result of the benefit could ultimately provide the incentive for the increase in price of low gravity crude in California. We are still producing our wells, receiving 50¢ per barrel less for our crude now than we received in 1957, and are faced with higher labor costs, material costs, and taxes.

It is most interesting to note that this one cent per gallon increase in gasoline prices brings the price of gasoline up to only 2% higher than the 1957 price level. On the other hand, labor, materials, supplies and commodities essential to our producing operations have increased at least 22% during this period. The general cost of living has increased by 15%.

It is with deep concern that we question the right of the Administration to threaten with reprisals the industry that has devoted 100 years to our consumers' needs and has been ready in every national emergency, fully prepared to serve our nation. Does our Secretary of Interior have the authority to threaten us with open-door imports of cheap, unreliable foreign crude, along with tax-exempt imports of gasoline? Does the Under-Secretary, Charles Luce, completely without knowledge in petroleum affairs, have the right to further blackmail a free enterprise system?

There may be other reasons besides threats and reprisals that will cause some of the majors not to join in the price hike. There is always the chance that some ambitious chief executive, regardless of his personal beliefs and standards, might accept a "special handout" for his company as compensation for "going to bed" with the bureaucrats.

In the best interest of the California producers, we ask you, as a Delegation or individually, to take the lead in getting appropriate committee action in Congress to launch a full scale investigation to determine by what authority Interior officials are exercising this control.

Yours very truly,

Johnny Mitchell
President

May 15, 1967

TO: Mr. John Doe
Washington, D.C.

Dear Mr. Doe:

I promised you in my last letter of December, 1965 that I would tell you how well the independents are doing under the Interior Secretary, Stewart Udall and our President, Lyndon Johnson.

I hate to say that our real problem can be traced down to our President, Lyndon Johnson. Especially since he is a Texan and knows more about oil than any other president.

I have to blame the President because of two things. First, when we had a chance of getting a good import law in 1958, when he was Senate Majority Leader, when we had a chance of getting the 16.6% limit of domestic production on imports and products placed into law so the bureaucrats couldn't twist it; Senator Johnson would back us only enough to get some mealy mouth words in the law which the bureaucrats would distort at will.

Then, when he became President he assigned the oil job to Secretary of the Interior, Stewart Udall, a man who is interested in nothing but Parks and Socialism . . . no matter what shape the country's security gets in.

I feel that the President is responsible even though he may not know how bad the situation is because he thinks Secretary Udall is in charge when Secretary Udall is thinking only of how he can please the President and his friends. You know, both the President and Udall are overlooking the real issue; how the oil industry is really doing, because they are looking at the profits of a few major companies and maybe a handful of individuals on the upper ring in the oil business, and they haven't faced the fact the real oil finders are falling out like flies.

As much as I hate criticizing President Johnson, I have to admit that he is truly the most knowledgeable individual on oil to even hold public office and has still yet to face up to the fact that our nation's energy security is deteriorating.

The President doesn't seem to care that he allowed a man like Commerce Secretary John Connors to open the gates of free trade zones for petrochemical companies while he was in public office and then, after opening up free trade zones and handing out quotas for petrochemical companies that would further demoralize the domestic industry; go to work as the head of a giant petrochemical company.

Why would the President permit Secretary Udall to hand out import quotas in Puerto Rico where his close friends, Able Fortas and Oscar Chapman were both involved. Why is he about to do the same with friends like Hess and Tommy the Cork in the Virgin Islands. The silly argument that these handouts are to resurge their economy is the flimsiest excuse that has ever been offered to the American public. Even the natives of those countries don't take that seriously.

I don't think I'll be writing you again soon. All I can see in my crystal ball is a steady deterioration of our domestic producing oil and gas industry. As we go down hill, imports of unreliable foreign oil continues to increase. It doesn't take a crystal ball to tell us the story that will unfold in the near not too distant future. The dramatic rise of the Middle East from a nation of camel drivers to become the wealthiest nations per capital in our present world. On any given day at their command they will bring every nation to its knees without firing one shot. All that they have to do is turn off the valves that furnish the oil that feeds all the energy hungry nations.

If you happen to see President Johnson, tell him that he still has time to dig into Secretary Stewart Udall's great give-away program. Remind him that he is still the President, and all of Udall's secret handouts will eventually fall on his shoulders.

Yours very truly,

Johnny Mitchell

October 26, 1967

TO: Honorable Stewart Udall
Secretary of the Interior
Washington, D.C.

Dear Mr. Secretary:

The decision of whether independent oilmen remain as an intergal part of the domestic industry depends upon immediate action on part of government acted on by your department. You must be realistic about the many problems confronting the domestic oil independent; problems which began a few years ago but have finally come to a head under your authority.

Your decision to open floodgates of foreign oil in such proportions as to threaten the future of the oil domestic producing industry has to be judged against you and our President Johnson. At the present rate of special handouts and now the establishment by you of special trade zones, imports are destined to reach a level about 30%. This continued increase in imports from the Middle East to special interest groups will some day make the Arabs block the masters of this troubled world.

Our Eastern Senators and people in government who are supposed to be knowledgeable, continue to spread the rumor that Middle East oil costs pennies per barrel. This is an illusion. This does not include the cost of our Mediterranean fleet, strategically placed and continually on maneuvers for the primary purpose of protecting these interests. It does not include the trained strategic troops that are trained especially for these same interests. It does not include the cost of building hospitals and schools, nor the cost of constructing highways, etc . . . things that the Middle East expects.

It does not include the tremendous damage to our own domestic industry, which has seen exploration virtually limited, the disappearance of thousands of domestic operators, the loss of experienced drilling contractors, the unemployment of thousands of field workers, and the other untold losses to our economy.

Never mentioned in the cost of oil is our balance of payment and growing loss of national security. Our greatest loss will be the day we awaken and discover that our nation's involvement with the Arab block in Middle Eastern oil has for many years helped to finance the Arab-Egyptian war machine against tiny Israel.

We are subjecting ourselves to international blackmail possibilities with our own current policies in the Middle East. It is quite conceivable that some day the Arab nations will force the United States to choose between the Middle East and Israel. With so much money and profit involved to a selected few international companies there is no doubt in my mind whose side the United States will choose.

We are in essence paying through the nose for foreign oil and waiting for the day when we must buckle down to Arab demands.

Sincerely,

Johnny Mitchell

July 8, 1968

TO: Honorable Ronald Reagan
Governor of California
Sacramento, California 95814

Dear Mr. Reagan:

We have never met, and by way of introduction, I served as the first president of the California Independent Producers & Royalty Owners Association in 1966, as President of the Texas Independent Producers & Royalty Owners Association in 1963-1964, and served as a member of the National Petroleum Council during President Kennedy's administration.

As a result, I feel I know the working attitude of the major integrated oil companies, both in California and in other producing areas of the country. What I am concerned about is the monopoly trends of these major integrated companies in their operations in the State of California. This to me is all more strange since the same majors don't exhibit this same trend in their operations in other areas of the United States.

I am, therefore, taking the liberty of asking that you take time from your important work to examine a problem which, no doubt, you would much prefer left to others. But the matter of artificially depressed crude oil prices in California does concern every citizen of our state, and I believe sincerely it is one which can only be dealt with effectively if the Governor of California personally demonstrates that he does not approve of the monopoly-like conspiracy which is depressing crude prices here.

For the past ten years, independent oil operators who operate in California have been trying their best to obtain some improvement in crude oil prices, and have repeatedly urged the few home-based major integrated oil companies which set prices on the West Coast to correct the chaotic situation.

The problem with California's oil producing industry is that major integrated oil companies operating on the West Coast pretend that we are insulated by geography from the United States insofar as oil and gas supply and pricing matters are concerned. Crude price increases in every area but California were initiated by some of the West Coast operating major companies, yet not one of these companies will assume the necessary leadership to apply the same industry value to our California crude.

The economic problem of the independent producer resulting from artificially depressed crude prices in California is a matter of concern to us, of course. But more important is that low crude prices are depriving the State of California of at least $500,000,000 in extra revenue in the East Wilmington pool alone.

The State's 85 percent net profits scheduled with low crude prices will take an additional three to four years before it comes into effect. In addition, the THUMS group has a continued use of this low price crude which will be refined into average priced products based on crude prices that have been restored in every area except California. Yet, these average priced products have no bearing on the net profits paid to the State of California. This gives this operating group a double take.

I charge that there is a monopoly situation in California participated in by these major integrated oil companies who control crude prices. Most of these companies within the past year and a half have restored crude prices in every producing area of the United States, except California, to a price per barrel higher than crude prices were in 1959. These same companies are paying California producers an average of 68¢ per barrel below 1959 prices and are not making any attempt to correct these prices. Unless these companies had a singular purpose, one or more with a need of crude supply would have stepped forward in increasing the price of California crude.

I sincerely hope you will not simply forward my letter to your Cabinet Secretary, or Director of Finance. Unless there is personal interest on your part, they will answer my letter somewhat as follows:

> "The State of California is deeply interested in obtaining a maximum return from its interests in the East Wilmington Field as well as from its manifold oil interests in many other tidelands areas in the state.
>
> As you may know, the price received by the State for its share of the crude oil produced from the field is based on receiving the highest arithmetical average price paid and posted for like crude oil of the same gravity in the Wilmington, Huntington Beach, Inglewood, and Signal Hill (Long Beach oil fields, whichever is the greatest. The State Lands Commission maintains a staff function in Long Beach to monitor and compare the prices paid for its share of the produced crude oil to assure the State of its receiving the best price available.
>
> The pricing of crude oil in California market—indeed the District V market—is certainly an important factor, both to producers and refiners alike. Crude oil pricing is not a unilateral function in itself, but is intimately involved with many important factors such as import quotas and production from other domestic areas. We can assure you that the problem is being considered in depth, hopefully with eventual positive results. The State of California, of course, has no legal authority to set the price of crude oil."

I already know, you see, what they feel can be said—in the absense of interest on your part. I would then have to answer along these lines.

> "It is true the State of California has no legal authority to set the price of crude oil, but it has the responsibility to investigate a monopoly."

Should you advise me that you will also forward my letter to the State Lands Commission for their investigation of this matter, I must tell you, Governor, that the State Lands Commission in the past have been too involved with major oil company executives to render impartial judgment. These major oil companies are quite wise in dividing the State Lands Commissioners among themselves so that each commissioner has only a few oil companies to favor. Since the monopoly conspiracy of these companies in California is single purpose—"low crude prices"—the State Lands Commission is not apt to apply pressure on their good hosts, the major oil companies, on this most important issue.

So, Governor, I suggest there is but one realistic course: you to tackle this issue personally. State revenues are your responsibility. If I may suggest, I would invite the following executives, collectively or individually. Here they are by name: Otto Miller, Chairman of the Board, Standard of California; Fred Hartley, President of Union Oil Company of California; Forrest Shumway, President of Signal Oil & Gas; Roland Eckis, President of Atlantic Richfield; George Getty, President of Getty Oil Company; Dick McCurdy, President of Shell Oil Company; Howard Ramblin, Chairman of Texaco; Rawleigh Warner, President of Mobil; Mike Wright, Chairman of Humble Oil & Refining Company. These gentlemen control California's oil industry.

We recognize that you do not have the authority in California to set the price of crude oil, but as Governor of California you must surely be aware of the millions of dollars these companies are pocketing that belong to the State of California. You could, in this conference, act in the best interest for the State of California. You hold the hammer in many other ways in which these company officials may find it quite difficult to operate and get approval on many urgent important projects in California. To permit and expedite these many projects may be the basis to warrant the adjustment of crude prices long overdue in California.

Each of these executives will seek to excuse their unstatesmanlike actions, and, of course, deny any conspiracy to depress California's crude prices for short range benefits to themselves and their companies. But there is not the slightest doubt in my mind that they will take time out from their selfish pursuits to review just where their attitudes and actions are taking us all if they are confronted by an expression of concern from the Governor of California.

Governor Reagan, I realize this is not the kind of request you would prefer to receive from a person who respects you, both personally and politically. But I sincerely believe it is in the best interest of the State, as well as the domestic independent oilmen. May I say that it is only because in you I see honor and integrity that I presume to request your personal intervention. Quite obviously, I would never have suggested that your predecessor do something so much in the public interest, because I personally do not believe former Governor Brown and so many politicians like him are capable of distinguishing the public interest from their own selfish political aims.

So long as major company executives are good political contributors, many in political life are unwilling to challenge their business practices, whatever the cost to principle and whatever sacrifice to State revenues. Such actions as this may not be the most popular thing you could do politically, but in my judgment it is the right thing to do and should be done.

Respectfully yours,

Johnny Mitchell

November 18, 1968

TO: Honorable Ronald Reagan
Governor of California
Sacramento, California 95814

Dear Governor Reagan:

In answer to my letter to you on July 8, 1968, I received a letter dated August 29, 1968, from your Director of Finance, Mr. C.W. Weinberger. Instead of answering him at this time, I prefer writing you again with a copy to Mr. Weinberger, which will save you the trouble of forwarding this letter to him.

The State Land Commission recently released information regarding the long-awaited, expensive, State-Paid DeGolyer and MacNaughton "Report on Crude Oil Pricing in California." The State Land Commission could have gone directly to Standard Oil of California Chairman Otto Miller for the same information, free of cost, for this is undoubtedly where DeGolyer and MacNaughton had to get most of the information for their report. This type of work is not DeGolyer and MacNaughton's cup of tea.

The underpriced crude in California has reached a critical point as it affects not only the survival of the small independents, but will deprive the people of California of millions of dollars in oil income from net profits operations in the East Wilmington Pool and other state-owned production offshore.

In the East Wilmington Pool alone, the State of California is to receive an average of 96½ percent net profits. Purposely agreed upon low crude prices among the majors makes certain that net profits will not come into effect until the peak economical producing capacity has been reached and passed. Net profits that were originally scheduled to come into effect in 1970 do not come into effect until late 1973 or early

1974. In the original projection of East Wilmington's expenditures and income, the operating group projected a 5 percent increase per year in labor and material costs. They also projected an increase in price of crude to offset the increase in labor and material. So far, labor and materials have increased by 20 percent, but, as a result of what seems to be mutual agreement among the majors crude prices have remained constant, thus extending the net profit period to the State by at least four years. This means the peak daily production of 160,000 barrels per day will have been reached and the high expense to maintain this production will occur at about the same time net profits to the State of California commence. Reliable sources state that the State will lose at least $500,000,000, based on low crude prices.

It is public knowledge that you are being advised not to get involved in any oil issues in the State of California. This is probably one of the reasons Mr. Weinberger asked me in his letter to go to the Justice Department in Washington if I had any question on monopoly trends of the major oil companies in California. This was a strange suggestion from a Republican Administration when a recognized monopoly in crude pricing is apparent only in California. More so, when only the State of California is being deprived of hundreds of millions of dollars. Of more concern to Mr. Weinberger should be the fact that these millions of dollars of the taxpayers' money winds up in the treasury of a few California-based major oil companies.

It is also public knowledge that you do have oil advisers. Unfortunately, this advice is major-oil-company-oriented. You also have one key adviser, a man of prominence in Los Angeles, who is close to the major oil companies and has been a servant of their needs for a great many years. It is hardly likely that his advice will be in opposition to the major oil companies and their unfair practices in California. His ties are too deep, his obligations too strong, and his knowledge of the independent oil man's problems too limited.

A few of us have recognized that, California, a leading oil producing state, doesn't have a State appointed or elected agency that governs inequities in oil and related matters, market demand proration, protection of correlative rights of individuals, or prevents waste of natural resources.

The DeGolyer and MacNaughton Report claims that imports are not a problem in California. Yet the demand for crude oil in California per day has reached a staggering 1,800,000 barrels per day. This means that we import into California approximately 800,000 barrels per day. This important fact was cleverly reported by the D & M Report to conceal the true ratio of imports.

The D & M Report states further that the supply of heavy fuel oil is increasing to such a point that in the near future it may have a more serious effect on future prices of low gravity crude. This situation is only aggravated by the blinded method in which the major companies are importing crude into California, denying the value of lower gravity crude.

The Report also claims that natural gas is replacing fuel oil in industrial use. The two-fuel economy is a competitive system that exists in almost every oil producing state.

As in the D & M Report, the major oil companies in California place the blame for low California crude prices on their import quota, the intrusion of natural gas, the success of steam flooding and finally, the East Wilmington Development.

In the past few years, all major companies have developed a hydrocracking or Isomax process which refines low gravity crude, recovering excess products out of each barrel refined. Under this new process, California low gravity crude should be priced at 65 cents per barrel more than they are now paying.

In addition to blaming imports for low crude prices, these major companies are now spreading the word around that when they recover the costs of these hydrocracking and Isomex units, they will make long overdue adjustments in crude prices. These units are merely a capital expenditure, similar to capital expenditures for thousands of new service stations and other capital requirements. It's only an excuse on their part, hoping you will buy it, Governor.

I would like to remind you that import quotas are being enjoyed in every producing area of the United States. On the East Coast and Gulf Coast, these same major California-Based companies have built hydrocracking and Isomax units at the same capital expenditure, yet crude prices in all other areas have been restored to the highs of 1957. California is the only exception.

If this is not a self-serving arrangement made possible by monopoly-type tactics, then I cannot imagine one.

Each President and Chairman of the Board of these major oil companies understand one code . . . . to serve only their company's welfare and that of their stockholders. Your personal advisers on oil are out to serve themselves and the major oil companies who have been their master for too many years.

However, you, Governor, have pledged to serve the State of California and to protect the State and the people's welfare to the best of your ability. I believe that is what you would like to do, else I would not bother to write you.

The State should have a selfish interest in California crude oil productions. The million barrels of crude produced per day in California pays approximately 10% of its value in taxes. This amounts to approximately $83,000,000.00 per year. On the other hand the 800,000 barrels of oil imported per day does not pay the State of California any taxes. For this reason alone, don't you think it would be wise, as other states do, to insure that our industry stays healthy and competitive and is not handicapped by the selfish, selfserving attitudes of the major oil companies of California.

Oil will continue to serve the State and its taxpayers faithfully for many years to come. The future of the small independent oilmen in California depends entirely on a fair adjustment of crude prices. Isn't it time your administration start fighting for a fair price for crude oil produced in California?

Only through your efforts can we expect a ray of hope.

Sincerely yours,

Johnny Mitchell

November 20, 1968

TO: Mr. Houston Flournoy
Chairman
State Land Commission
State of California
Sacramento, California

Dear Mr. Flournoy:

I have recently reviewed the full text of the DeGolyer and MacNaughton Report on California crude prices which was authorized by the State Land Commission at considerable cost to the State of California.

I must compliment the State Land Commission for being consistent. Consistent, that is, in wasting the taxpayers' money.

To prove my point, I will take you back to 1963 when there was a bitter struggle in the State Land Commission as to the ownership of the East Wilmington Unit. This was a bitter battle between a Pauley-dominated State Land Commission and the City of Long Beach. At that time, Senator Virgil O'Sullivan's watchdog committee, with State Land Commission approval, hired former Secretary of the Interior under Truman, Oscar Chapman, to make a report on the East Wilmington dispute. The Commission was aware that Oscar Chapman was an old crony of Ed Pauley.

Bringing in Oscar Chapman to review this problem was as impractical as having DeGolyer and MacNaughton report on crude prices in California. Particularly while under the gun of Republican advisors to the Governor. Advisors who are, incidentally, officials, current and retired, of major companies that control crude prices in California.

Oscar Chapman's report was delivered at night and he left California with his check for $30,000.00 under the cover of darkness. The DeGolyer and MacNaughton Report should have shared the same fate.

Respectfully yours,

Johnny Mitchell

February 5, 1969

TO: Honorable Ronald Reagan
Governor of California
Sacramento, California

Dear Governor Reagan:

I am taking the liberty of writing you once again in regard to crude oil pricing in California. I appreciated hearing from your State Finance Chairman, Mr. Weinberger, regarding the State's concern and investigation of this critical problem.

Crude pricing is normally the end result of supply and demand, plus good, sound marketing practices. These import factors, unfortunately, are neither sponsored nor practiced by our California-based major oil companies.

My correspondence with you has never been out of disrespect for you or your office, but solely in defense of a troubled, independent segment of our industry. I respect you as a Governor and as a man.

We independents have exhausted our search in California for leadership and statesmanship. In a last desperate effort, we hope that through your concern, a statesman within the major company ranks might emerge. The timber is there . . . . in such leaders as Miller of Standard, Hartley of Union, Shumway of Signal, Anderson of Atlantic Richfield, McCurdy of Shell, Rambin of Texaco, or Wright of Humble.

It doesn't take much to save the small, independent oil producer. Only an overdue, few cents per barrel increase in the price of low gravity crude. This is not too much to ask of any industry leader, especially when history has proven that the independent oil man has been the balance, the backbone, and the oilfinder of our industry for over 100 years.

In concluding my last appeal to you, Governor, I thought my letter of July 9, 1963, to Mr. R.G. Follis, Chairman of the Board of Standard of California, might be of interest to you. It was evident then, as it is today, the urgent need of a crude price adjustment and long over due industry leadership.

Yours very truly,

Johnny Mitchell

November 7, 1969

TO: Mr. Michael Haider, Chairman
Standard Oil Company of New Jersey
New York, New York

Dear Mr. Haider:

When Standard's Mr. Jack Rathbone retired, I felt that his departure from the industry would leave a vacuum in our relations with major companies that could never be filled.

I was wrong because as you succeeded Mr. Rathbone as Chairman some years ago, you proceeded to fill the vacuum up to the brim with the same understanding and compassion for the well being of the small independents.

It was never in our hopes that either of you could stop the downward trend facing the independents. Imports were continually increasing and regardless of your concern, it was just one voice crying out in a wilderness of greed.

I can't begin to tell you the assurance you gave me by your invitations at various times to discuss these problems. Each time that I would leave your presence, the urge to see other executives of other major companies became an obsession. However, the majority of other company leaders weren't a Michael Haider or a Jack Rathbone.

Now that you are retiring I pray that it will be filled with memories of your great adventure as you climbed the ladder starting at the bottom and reaching the top rung.

The pleasure in being able to know you and Mr. Rathbone has been a special high point of my life. I have locked up in my memories three great industry leaders, all belonging to the same Standard of Jersey family. They are Carl Reistle, Jack Rathbone, and Michael Haider. In the history of Standard to come, this combination will never be duplicated.

With the greatest respect, I'll end this letter and pray that the best years of your life lie ahead.

Sincerely,

Johnny Mitchell

December 4, 1969

TO: Johnny Mitchell

FROM: Mr. Michael L. Haider
One Rockefeller Plaza
New York, N.Y. 10020

Dear Johnny:

I appreciated very much your letter of November 7th and your kind remarks in connection with my retirement. It was good of you to write. I am sorry to have been so long in responding, but have taken some time off at my place in the country and have only this week returned to New York.

The petroleum industry certainly is faced with many problems these days and I am not at all happy as to the trend things are taking. However, I feel the industry has made the best possible representations to the appropriate people in Washington and, from this point of view, has done everything possible. I think our real problem is the lack of understanding of the petroleum industry by the public at large. At the last meeting of the API Board, we put in motion some steps to develop such programs which we hope will correct this situation.

Again, it was good hearing from you and I hope that our paths will cross before too long.

Sincerely,

M.L. Haider

December 2, 1969

TO: The Honorable Wilbur Mills
House Office Building
Washington, D.C. 20515

Dear Congressman Mills:

May I make one last appeal to you as a key member in formulating the final draft of the present tax reform bill.

Having had the privilege of discussing with you the merits of various tax proposals levied against the oil and gas industry, I believe some could be modified to fit into the tax reform without damaging this vital industry which is so essential to our national security.

We both recognize that as far as the independent oilman and his small investor are concerned, the most essential tax incentive that will keep us able to continue to explore for needed reserves is the intangible expensing feature of our tax laws. Of lesser importance is depletion.

We were heartened with your committee's action on intangibles and we felt we could live with the adjustment in depletion. As I have said before, what good is depletion if we have no oil to deplete?

The Senate Finance Committee, in its final version of the Tax Reform Bill, included the intangible expensing in a modified form in limiting so-called tax preferences.

Unless intangible expensing is taken out of the limited tax preference, the small independents and their investors who, in the past have found 85% of the oil and gas reserves in America, will be unable to continue finding the necessary reserves that are so vital to our security.

We are unable to understand why the Senate Finance Committee put the intangible expensing back into the limited tax preference treatment when there are other features that could generate more tax income and would do less harm to the industry's future development and exploration program.

It is our hope that the conference committee will recognize that depletion adjustments are far more acceptable to the independents (as well as adjustments on foreign tax credits, ABC carve-outs, and foreign depletion) than any provision limiting the privilege of expensing intangibles.

We independents, with deep concern, call for wisdom and responsibility in this issue and for the joint conference committee to recognize the damage that the intangible expensing in the limited tax preference treatment will have on the independent segment of our industry.

Respectfully yours,

Johnny Mitchell

December 8, 1969

TO: Mr. Johnny Mitchell

FROM: The Honorable George Bush
The United States House of Representatives
Washington, D.C.

Dear Johnny:

Thanks for sharing with me your letter to Congressman Mills.

In case it escaped you, I want you to know that I was the one who introduced the motion in the Ways and Means Committee to remove depletion and intangibles from the limited tax preference formula.

This motion carried; but, unfortunately, the Senate through their minimum tax provision undid the work that we did in the Ways and Means Committee on intangibles. The depletion provision in the Senate bill, of course, is better than the House bill.

Your letter to Chairman Mills was a good one.

Best regards,

George Bush, M.C.

December 9, 1969

TO: Mr. Johnny Mitchell

FROM: The Honorable Olin E. Teague
The United States House of Representatives
Washington, D.C.

Dear Johnny:

I have just received your letter of December 2nd. I have talked with Wilbur Mills about both letters and hope it will do some good.

It is always good to see you—I expect to be in Houston for a couple of days the first week in January and may see you then.

Sincerely,

Olin E. Teague

May 7, 1970

TO: Mr. Clinton Engstrand, Chairman
Liaison Committee of Cooperating
Oil & Gas Association
719 Union Center
Wichita, Kansas 67202

Dear Clint:

I would like to offer a few suggestions that you might consider discussing at the Liaison meeting in Denver on May 10. As you know, I have always felt that an import quota for the domestic producer was the only incentive that could keep us competitive with major international oil companies. Our fellow independents are a stubborn bunch. False pride has taken a heavy toll of our ranks because of our unwillingness to recognize the necessity of a producer import quota. It has finally soaked in that a quota is not a subsidy, but an essential ingredient for survival.

Many feel that my last speech on tariff was an indication of my approval of Secretary Schult's tariff proposal. To the contrary, tariff is our last hope, if producer quotas are not included in the proposed import program. The purpose of the speech was an attempt to tell the industry that anything less than a equal-sharing import program would not be as advantageous to the independent as a tariff.

The new program must eliminate all inequities, historical quotas, Puerto Rican and Caribbean handouts, trade zone exemptions, Canadian control and other exemptions. To continue with a patch-work formula is merely an extension of the present, unworkable program.

Secretary of Interior Hickel proposed an import program. His won't work. Not only has he added new problems, but he is recommending all the bad points of the present program. He failed to eliminate the 200,000 plus barrels of handout, exempted imports going into Puerto Rico and the Virgin Islands. To make matters worse, he continues to allow 75,000 barrels per day of duty free gasoline, refined from this crude, to enter all market areas on the East Coast, West Coast and Gulf Coast. He should disclose that cut-throat marketers can purchase gasoline from Puerto Rico for any port in the U.S. at a price per gallon always a few cents under the domestic market price. This disadvantage not only riddles fair marketing practices, but the present program as well.

Along with many other loopholes, Secretary Hickel also insists on keeping the historical quotas that should have been phased out in 1963.

However, if the Administration elects to keep the existing program we must make a concentrated effort to be included.

From all indication the appointed IPAA subcommittee is making a weak effort in agreeing on the merits or a producer quota. We have always leaned toward an incentive quota based on total drilling.

However, out of the subcommittee meeting last week Bill Cleary of OIPA, recommended an exploratory dry hole incentive quota. We will buy his idea, so get behind Bill Cleary.

Our error has been in letting IPAA drag its feet on producer quotas. Let's face it. Their import committee is dominated by importing companies. If IPAA doesn't come forward with a constructive producer quota, you have one chore left as leader of your Liaison troops. March the real independents out of the meeting and leave the rest behind who have been hiding in the Trojan horse all these years.

Sincerely,

Johnny Mitchell

June 16, 1970

TO: Honorable Olin E. Teague
House Office Building
Washington, D.C. 20515

Dear Congressman:

Relative to our telephone conversation a few days ago, we will be in Washington on July 16, 1970 and would appreciate it very much if you would call Del Brockett of Gulf Oil Corporation and set up a meeting in your office with Del Brockett, Otto Miller, Chairman of the Board of Standard of California, Ken Jamieson, Chairman of Standard of New Jersey, John Swearingen, Chairman of Standard of Indiana, Robert Dunlop, Director of Sun Oil Company and any other major oil company officials that Del might suggest.

Frank Ikard, President of API, advised that he would do all he could, cooperating with you to the fullest extent, to help call this meeting. By now Frank has probably reached you by telephone.

We would appreciate it further if you would select a few Texas Congressmen to join us along with Congressman George Bush and Senator Dole of Kansas.

The meeting is not intended to embarrass the major oil companies, but to acquaint them with the critical situation in which we independent oil men find ourselves. I know that if we continue under the present competitive conditions and share no part of the import program it is enevitable that almost every independent will be out of business. It is my opinion that individually, these major company officials are unable to lean toward any policy that will help our situation, but an appeal to them as a group might open the doors to administrative action that would relieve our situation.

I find it hard to believe that after all these years everyone shares in the import quota but the independents who have found over 75% of all the oil and gas in this country. Our position today makes it prohibitive to even consider exploration or development without an incentive.

This is only a brief message, but I think you might understand what I am talking about.

I am enclosing a copy of a speech I made in February that brings out several points you may not be familiar with. I hope you will review it and if you have any questions, we can discuss them.

It would be difficult to express my appreciation, and that of the other independents that are aware of the situation, for taking our problems on your shoulders.

Sincerely,

Johnny Mitchell

September 7,1970

TO: Honorable Emilio Q. Daddario
House Office Building
Washington, D.C.

Dear Congressman Daddario:

Since a remark by you has drawn a sharp response by a former assistant Interior Secretary, J. Cordell Moore, in the August 3, issue of The Oil & Gas Journal, I believe it wouldn't be out of order for me to direct a few words your way.

It was my happy privilege to meet you by accident in the office of one of my good friends from the Texas Delegation in Congress, and I was deeply impressed with what seemed to me to be a Congressman from Connecticut who is willing to make a statesman-like effort to understand what's really going on in the oil industry today.

This reinforced my belief that most of the misunderstanding between the oil & gas producing states and the New England "Consumer" states arises out of the breakdown in communications. I believe this is true insofar as the domestic independent producer is concerned—as distinguished from the major international oil companies whose interests may differ from the interests of the people you represent. Maybe it's timely for consumers in the Northeast to try to understand that there are two segments in this petroleum industry, the majors and the independents.

As you will recall, the point I was attempting to make to my Texas friends was that it is high time the import program was changed. Now is the time to allow the independent producers and explorers some kind of direct participation in the benefits of the quota allocation system.

This nation is currently producing all the oil it can produce on a sustainable basis, and is living off the shelf so far as the future is concerned unless there is restoration of former drilling levels. Today we are drilling about one half as many wells as we were when the import program was put into effect, and continuing to lose ground. Why? Because the import program doesn't do what it is suppose to do—and neither the Administration nor Congress seems to want to face up to that.

We covered the reasons behind the failure of our industry's effort in maintaining this nation's energy requirements such as, the true role major oil companies have played these past eleven years in selling out this nation's domestic development in preference to developing unreliable foreign sources. The responsibility of government in allowing major companies to hog all the incentive import quotas these past eleven years, assured them added subsidies for financing their foreign operation.

It wasn't hard to explain, as the result of economic manipulation by major companies, this great nation could degenerate from the world's leading energy producing nation to one dependent upon unreliable foreign sources for our daily energy requirements.

After years of running blind we finally awaken and discover we possess none of the reserve energy requirements necessary to sustain us as a first rate world power. Can you believe, Congressman, we are fast approaching the stages of a second rate power, critically short of electrical and other energy, primarily because we have a Petroleum shortage?

Can anyone really defend the actions of government and major oil companies these past eleven years when people must turn off their air condition units in the summer and maybe do without heat in the winter?

Where do you suppose all the incentive of import quota went . . . into domestic exploration as it was intended? No! Some of it went to finance, in part, the major companies' development of all the useless foreign energy sources we are now writing off as unreliable. But most of the import program advantages went into added profits to the majors.

For eleven years, the import quòtas have been denied the independent segment as an incentive to find new oil. Where is the logic of this?

Personally I think it is incumbent upon these persons responsible for governmental oil policies, especially Secretary of Interior Hickel and General Lincoln, to find the key to the import quota program that will unlock the incentive for domestic exploration. The major oil companies who have been receiving the import quota these past years have failed miserably.

Let's try a new formula . . . . let the domestic independents go to work.

I believe you are one member of Congress from New England who understands the essence of what I am trying to say. In turn, I believe we independent oilmen will try a little harder to understand consumer needs, and never oppose reliable imports of oil from any source that is really necessary to assure the energy needs of our great nation.

I am taking the liberty to send a copy of this letter to people who might be interested because this delicate subject "Energy Crisis" is of increasing concern to our nation.

Respectfully yours,

Johnny Mitchell

July 24, 1971

TO: Mr. H.J. Haynes, President
Standard Oil Co. of California
225 Bush Street
San Francisco, California 94104

Dear Bill:

Over these many years, I've watched industry leaders climb the ladder to the top as if they were handpicked by Fate. As each reached the top, there was always our hope that at least one would detour from the competitive scramble for profit just long enough to see how badly the industry is faring, especially for the independents in California.

The room at the top has always been crowded with outstanding executives, usually breaking bread over their company's successes.

Each will tell you in private the same words, "I remember the early days of this industry, the helping hand of the major oil companies toward an independent." I would really like to take time to review the many problems and help the small independents." "If I do the other international giants will race right by me." "I can't ask for a reduction of imports for my company, for if I do another will get it."

Still Bill, with this competitive attitude now shared by all, the end result will be that doomsday will be here before you retire. And like the rest, you will be looking back and wondering why your company didn't help the small independent in their struggle for survival, and then you will fold your tent and like the leaders before you, silently slip off into the darkness of the night.

As we look beyond our present problems here in California and, at the bleak future of supplying this nation's energy requirements and review the recent statements by President Nixon that government is willing to spend millions of dollars to ensure our future energy requirements, we can only wonder how this nation failed to develop its oil and gas reserves. Especially when just a few short years ago our oil and gas reserves far exceeded our energy demands.

Some argue it serves no purpose to criticize further the past hopeless policies, or any parties responsible for the serious exploration slowdown of the industry. Instead, this should be the time to face facts and devise new policies that will assure the domestic industry of not only adequate reserves, but a reserve producing capacity that we can depend on without fail.

However, I disagree. It's important to assess blame. And for my part, I say it is the major companies which deserve most of the blame. One cannot brush aside the past ten years that saw so many independents go out of business. So many appeals to both major oil companies and government for a share of the import program, so many other appeals for cooperation for a chance to share a few incentives that could have kept the small independents alive—all falling on deaf ears—ignored by both.

Executives will seek to excuse their unstatesmanlike actions and, of course, deny any intent to jeopardize the nation's energy adequacy. They are guilty of failing to look beyond short range benefits to themselves and their companies. There is not the slightest doubt in my mind that they will never take time out from their selfish pursuits to review just where they failed their industry and their nation—unless we help the public understand how we got in this mess and who is responsible.

So long as major company executives enjoyed such good relations with federal oil policy makers, how could we expect political life to challenge their business practices, whatever the cost to principle and whatever sacrifice to the nation's security? Maybe it's still hopeless. Challenging the policies of major oil companies still may not be the most popular thing to do politically, but in my judgement it is no longer just the right thing to do, but has become the only hope for a better tomorrow.

Whether it is too late for the independent segment of this industry to regain its role, I don't know. Incentives depend on future understanding between the independent segment and the major international oil companies, even more than on better "public" or "governmental" relations.

We recognize that some bureaucrats will never acknowledge that crude must have a better price, the industry a better tax, and an import incentive to restore domestic exploration and drilling. They don't think it's politically safe to admit what they know is a fact. Even to the point that our nation is running out of energy.

It's high time some industry statesmen use the influence which is theirs to use. If they know it's right to raise wellhead prices of crude to get some more wells drilled at home, to get the independents back in business, to head off a disastrous energy crisis and possibly complete Federal Control of our energy industries—then they should do it. And if White House economists and White House Advisors and New England demagogues and a lot of self-styled consumer champions come to face up to the fact that the voices of the radical element among environmentalists and ecologists and their words of doom should be stilled while more realists and concerned appraisal of our energy needs should take the forefront.

At best, it's a race for our lives to see whether this great nation goes under for lack of energy.

At the present rate, we are heading for a crisis that will shake the world. When that happens, there no longer will be any priorities from government for Standard of California, nor for Jersey Standards or the rest of the majors. You will be joining us independents as we slide down the tube together. And most tragic of all, our nation's dependency on foreign oil will become a true way of life here in America.

Sincerely,

Johnny Mitchell

February 10, 1972

TO: Mr. Peter Flanagan
Executive Office of the President
Washington, D.C. 20504

Dear Peter:

The oil industry is nothing more than the sum of the individuals who make it their way of life.

The nation depended on these individuals. Once in abundance but through people misunderstanding and government controls these ranks have thinned to only a handful.

Under present policies the petroleum industry will not turn upward, and certainly won't sweep anybody anywhere.

A few individuals will succeed, others will fail, and the sum of their limited successes minus their failures, will be summarized as the nations progress in our current energy crisis.

We're not going anywhere in finding new oil and gas reserves and our nations energy needs are increasing. That's why I am writing you.

Through your office there is a chance that the domestic oil and gas industry may gain one of the key incentives necessary to increase the nations domestic drilling.

Price of crude oil, the key incentive, lies in the hands of the major international oil companies.

A producer import quota is the second key incentive of importance that is necessary to increase exploration. Through your interest, we may receive the necessary attention.

It's time to begin relating oil import policy more directly to domestic drilling, if we are to survive as a nation. An import policy which is the "welfare state" in reverse, where the big prosperous international companies get it all, can't work. It has run its course.

The oil import program will never work until it relates more directly to the original fundamental objective of keeping adequate producibility in America.

Leaders who readily acknowledge that the program has to be counted a failure still remain unwilling to accept changes which would allow some of the economic advantages if importing is to be siphoned off to stimulate domestic exploration and drilling.

I am not talking just about the government leaders within the Administration. And I'm not being critical of the Nixon Administration any more than its predecessors. There is blame enough to go around, and to include our Congressional friends. One Congressional leader himself, though he personally has done more than his part, recently acknowledged that the oil industry gets hit ten times by its critics from the "consumer" regions of New England for every time an oil-state Congressman stands up to tell the real story. Why? Because, they say, to speak out on oil gets them accused of representing special interests, whereas anti-oil demagogues can talk all kinds of nonsense and somehow be credited with statesmanship for standing up for the consumer.

Oil Policy chairman George Lincoln makes sense when on the broad issue of the energy crisis. But ask him about some form of drilling incentives and he seems to tune out. He seems most persuasive when blaming others for the crisis.

Shortly after he took over as Oil Policy chief, General Lincoln declared some major changes were going to be made in the oil import program. Instead he has continued down the same old path, to a point where we have now run out of reserve capacity necessary to our security. Even yet, he seems unwilling to move on clear-cut proposals for changing the import program which would both increase drilling this country and provide incentives for refinery construction in this country.

We are abandoning every year thousands of wells which could go on producing for years if the price of crude were even nominally increased. Not even Washington planners can make free citizens conserve that which costs more to conserve than it sells for. New conservation procedures without adequate incentives will prove little help unless and until realistic wellhead prices are permitted. And it's Washington policy which controls our crude prices.

In summary, while everybody in industry and government who has really looked into the situation acknowledges that America is in trouble energy-wise, nobody in a position to act is really doing anything. America stands on the brink of becoming a third-rate power for lack of energy—and neither the government nor the industry is acting to avert that catastrophe.

Now I don't want anybody assuming that I am accusing our leaders of being indifferent to what happens to America. They are good Americans. But I say they are guilty of taking the path of least resistance—with the ultimate consequence of reducing America to pathetic dependence upon unreliable foreign sources for its life-giving energy. Because of inadequate incentive, and refusal to make the import program into a vehicle for ensuring drilling in this country, we are having to increase imports everyday. What worries observers most is that the lion's share of this increase is having to come from the unstable Mideast, where Russia clearly dominates.

Without intending to single out any particular company, I do recall that only a decade ago the chairman of Jersey Standard testified that his company had no intention of bringing Mideast oil here because he thought the U.S. was not a logical market for it. Yet by standing with other companies in failing to do anything to keep America in a position to do without that oil, Jersey Standard must stand indicted along with others in bringing America to the position of having to rely upon Mideast for more than half of its oil in the early future. No doubt these large companies never intended to render America dependent upon such unreliable sources. Yet by pursuing their short-range interest at whatever expense to broader national goals, they have helped make this inevitable.

I can say without any hesitation that if this present administration gets behind a sound producer quota, it will be taking the first step in the right direction. A great many oil men believe it makes sense. Secretary of the Interior and General Lincoln don't.

Yours very truly,

Johnny Mitchell

June 10, 1972

TO: Honorable George Wilson
Executive Office of the President
Washington, D.C. 20504

Dear Mr. Wilson:

Through your office there is a chance that the domestic oil and gas industry may gain two key incentives necessary to increase our nations' domestic drilling.

The primary incentive is an increase in the price of crude. The second is a share of the current imports. There is no doubt that you clearly recognize that our National Security—our balance of payments,—our very survival—all require doing something about rebuilding our domestic energy potential. But there is a tremendous gap between the Nixon Administration's rhetorics and their actions. Talking great but doing nothing won't get the job done.

Everybody seems willing these days to talk a lot about what increasingly is recognized to be a genuine energy crisis. Top spokesmen for the Administration, leaders in Congress, and our so-called industry statesmen all give evidence that they know the problem is real enough—and some are now willing to declare it an emergency or near-emergency.

But there remains little or no hard evidence that the energy deficit problem is going to be attacked in ways—or in time—to lessen the catastrophic effects upon our economy and perhaps even our security.

You and The White House apparently think it's first necessary to get the public in agreement before doing what it knows is right. You said "If a national crisis is going to face in the not-too distant future, then we'll have to educate people to the danger, find and propose a solution and explain why it is in their best interest." This is a cop out, refusing to do something that is overdue merely because it's not yet good politics.

Our industry leadership is embodied best perhaps in what is known as the National Petroleum Council. This is a large body of leading industry executives, including a smattering of independents, who serve in an advisory capacity to the government. No effort is spared to keep this group in the forefront of industry events, and certainly they have at their disposal whatever resources they find necessary to launch whatever they find necessary. So it can be said the industry has through the NPC ample opportunity to cause the government and the public to face the realities of the situation.

But because of cross currents of special interests among industry segments and leading companies, the NPC has failed dismally to use its beautiful opportunity to register the case for a realistic national energy policy.

Take imports for example. In the years following World War II when oil imports became excessive, because it was good economics for the international companies to bring in cheaper foreign oil, the NPC stood paralyzed. Long after excessive imports had begun to erode domestic exploratory efforts, the NPC proved unwilling to face the problem—because most of its members were enjoying the full benefits of virtually unlimited cheap oil from abroad. When the problem could no longer be ignored, the NPC labored long and hard on import policy—and gave birth to a mouse. It reached the historic proclamation that "imports should supplement, not supplant" domestic production. Because this could be interpreted any way one chose, individual importers kept right on supplanting domestic production—until the Texas allowables were down to 8 days per month. Of course drilling suffered, and after a dozen years of this kind of abuse our defense-vital reserve productive capacity was finally depleted. Today we are on 100% allowable, actually producing in excess of 100% of the true efficient capacity without incurring waste.

Only recently have our industry statesmen permitted the true story of reserve productive capacity to be told in official documents. Long after they knew better, they were publishing figures showing the U.S. had well in excess of a million barrels daily unused capacity. The only thing resembling an excuse I've found for this tomfoolery is that they thought the OPEC nations were stupid enough to believe the U.S. could retaliate should Arab countries shut off our supplies or hike the prices too sharply. But we were only fooling ourselves. I suspect the Arab potentates knew more about the factual situation than certain leading brass of American companies.

In the process of trying to hide the truth, we brought on ourselves additional misery in the form of a wide credibility gap. When we finally had to tell the public the harsh reality that there is no reserve capacity, the retort was that this "shortage is a myth" designed to jack up the prices needlessly. We are still having to confront that nonsense in most quarters.

Well, you say, even if the leading voices of our industry were unwilling to face facts, that doesn't excuse a responsible government for not doing what its own studies showed necessary to avert an energy collapse in this nation. I would agree, except for the fact that in practice government leaders, in Congress and the Administration, traditionally take the assessments of leaders within an industry. And politicians are certainly no braver than industry leaders, especially while it remained good politics to pretend that the shortage would go away.

If industry leaders never bring themselves to looking above their own company interests, it doesn't excuse government leaders from doing something besides talking about the energy crisis.

Yours sincerely,

Johnny Mitchell

April 30, 1973

TO: Mr. D.F. (Jack) McKeithan, Jr.
Chairman, Liaison Committee
735 North Water Street
Milwaukee, Wisconsin 53202

Dear Jack:

In a matter of a few days we will all be descending on Phoenix, hoping for miracles that will never happen.

It seems clear that neither government nor major internationals are going to make our bed any easier. To gain an incentive, we will have to force it ourselves.

There are just three basic incentive sources: price, taxes, and import quota.

Unless we can do something to juggle supplies a little better, pricing of our crude and gas lies with the majors and the government. Even in a seller's market, it will be hard to budge either. The government under Nixon, judging from his recent message, doesn't want to recognize the domestic producing segment.

It may turn out that an import quota is the only incentive we can gain right now, and that will be only by a concentrated united effort on our part. Many say it's too late. I don't agree. No matter what they pretend, foreign oil is cheaper to majors. So I believe an import quota will continue to be worth approximately 35¢ a barrel in the foreseeable future. The government has to keep some value in them for its own reasons. This incentive, with a nominal crude price hike, will help us meet the sheriff half way.

The independents have fumbled the ball badly over these past 10 years in not taking advantage of the import quota. Majors are smarter. They tell us to be too "proud" to take the subsidies they take anxiously. They figure their survival on dollars and cents, regardless of where it comes from. Furthermore, the majors, in hogging all the import quota these past 12 years, use these millions of dollars of benefits derived from the quota to help finance their foreign operations—adding to our troubles by subsidizing imports indirectly.

The real guilt of these same majors is that while they lined their purses with profits of the import program, they turned their backs on domestic exploration and development. In 12 short years they selfishly watched this great nation slip from a leading power to become a second-rate energy beggar.

The downfall of the independent producing industry will never be cataloged. Most of the blame is due to our own apathy and false "pride" of independence.

I'm sorry to conclude that another contributing factor is the failure of "The Liaison Committee" to live up to its potential. Last but not least is the infiltration of some of our leading "independent" associations by importing members and their handpicked stooges.

IPAA took 12 years to come around to agreeing to give some benefits of the "import program" to the independents. Twelve long years of this, combined with the ineptness of the independents and the Liaison Committee hurt us badly. Over these years we watched the feeble attempts by independents and by Liaison to open the door of the import committee for the independents. They both failed because our unity of purpose was never modeled. You simply cannot be an independent association expecting to win the high ground unless you fight for the independent all the way.

It takes gutty leadership—we can't win by keeping on extending olive branches and pulling back bloody knuckles.

We need an import quota based on domestic exploration and development. It *will* always be worth something. For the first time, it should be easy to put over in Phoenix. The Import Committee of IPAA could care less, the flood gates are not open. Let's take advantage of an opportunity even if it is late.

Yours very truly,

Johnny Mitchell

October 31, 1973

TO: Honorable Henry Jackson
Senate Office Building
Washington, D.C. 20510

Dear Senator Jackson

I am taking the liberty of writing you about a critical situation that is already present and will only worsen as time goes.

For the first time in twenty years the price of crude oil and gas is expected to reach a price that will rejuvenate domestic exploration and development of oil and new reserves.

Now that the independents can forecast a realistic price, we have increased both our exploration and development activities. But now, unexpectedly, we face a new serious problem: a serious shortage of tubular goods. This alone can put the brakes on oil and gas exploration programs vital to our nation.

Independents don't stockpile tubular goods for the simple reason that they do not have the capital. Independents, who actually drill three-fourths of all domestic wells, acquire their tubular goods from supply houses as they drill, at the time they are actually drilling, and require casing and tubing to complete their wells.

As matters stand today an independent cannot begin a well unless he has a firm commitment of casing and tubing. Up until six months ago supply houses kept an adequate inventory of casing and tubing to take care of the independents. These same supply houses today report they are completely out.

What caused this sudden shortage? The basic problem is inadequate capacity to supply steel demand. But the immediate cause is more susceptible of short-range solution. It is a run on the mills by a few large companies—stockpiling against future needs. By so doing these major companies are denying independents the opportunity to expand exploration and drilling.

To put it simply, something needs to be done to keep any company from acquiring and sitting on tubular goods which it has no immediate plans to use.

I am not pointing the finger at any company for stockpiling, or charging that any major mill has favored these companies. In normal times, the right for mills to sell to whomever they please should be their privilege. Yet in the face of our severe energy crisis, when our nation requires every barrel of oil and every cubic foot of gas we can produce, I feel that any company in the national interest should be willing to put part of its excess back into stock to allow the drilling of more wells now. I also believe that major mills would be agreeable to some form of allocation and distribution procedures, provided that the urgency of the situation is called to their attention by someone who is concerned with our nations' welfare.

For this reason, I am calling on you to review the current situation with these suggestions:

(1) Require all major pipe mills to report their mill capacity; their distribution to major suppliers; their direct distribution to all oil companies providing a complete record of all 1973 sales and to whom.

(2) Require all oilfield supply houses to supply same information.

(3) Require all companies and gas transmission companies to report their oilfield tubular goods. This must include all their subsidiaries and affiliated operations.

(4) Require all companies to limit inventories to a six-month requirement of tubular goods based on definite drilling projections.

(5) All excess be set aside for redistribution to supply houses for independent domestic drilling requirements.

(6) Provide some form of limitation on all exporting of tubular goods that will only benefit unreliable foreign oil development, especially where foreign pipe is available.

(7) Set up some form of allocation program with major pipe mills so that sufficient mill capacity is earmarked to supply houses for current domestic exploration.

(8) Take steps to change the Cost of Living Council regulations and other impediments to importation of steel as needed for domestic energy development.

I respectfully submit that measures along these line would go far to alleviate this serious situation and serve a vital national interest.

Respectfully yours,

Johnny Mitchell

January 2, 1974

TO: Mr. Robert O. Anderson, Chairman
Atlantic Richfield Company
1515 South Flower Street
Los Angeles, California 90071

Dear Bob:

When you call for elimination of percentage depletion allowance, you are also speaking for the independent producer who has accounted for some three-fourths of the oil and gas found in this nation. So let the record show that you do not speak for the typical domestic independent oil and gas explorer.

It very well may be that you and Arco President Thornton Bradshaw have figured out some way Arco could gain competitive advantage over other major petroleum companies through this maneuver. Maybe Arco is embarrassed by its earnings picture, doesn't need improvement in its earnings in order to carry its share of the load in making available more energy resources for our nation. But domestic independent producers, unlike the international giants, have not been enjoying record earning and growth in recent years.

As a matter of fact, because in large part of policies dictated by international companies, we have undergone a very serious depression throughout the past fifteen years. Drilling in this country was cut in half—with independents suffering the entire load—from the level of the mid-1950s. Major companies suffered no part of this cutback, because they continued throughout to drill about as many domestic wells as they had, while enjoying relatively unimpeded privilege of importing their then-cheap foreign holdings. Because foreign oil was so cheap during this period, the major companies held domestic crude prices well below replacement levels. The predictable result is the current energy crisis—a crisis which could have been avoided by this nation had independents not been virtually squeezed out of the business of searching for new oil and gas.

If Arco wishes to forego percentage depletion on its U.S. or foreign operations, there is absolutely no prohibition of it under present law. I trust you will have the courage of your conviction and volunteer your company for this test of how effectively it may serve without percentage depletion.

I realize you can't be serious, of course. But the damage to our industry could be immense from this kind of childish behavior. It is one thing for this suggestion to come from someone in a position to assure us a free market price for our product in exchange for our yielding up percentage depletion. It is quite another for executives of a leading international company to do so. All you can possibly hope to do is confirm the erroneous impression of our industry's professional critics that depletion is unnecessary and unjustified.

What irks some of us most is your willingness to give up what is most vital to the nation's real oil and gas finders, something of relatively less importance to international companies. As disgusting as it is, this is not altogether surprising to those of us who have witnessed the lack of statesmanship exhibited throughout the past decade-and-a-half by major company executives.

Now I am perfectly aware that we are under siege as an industry. Even our President recently echoed the critics' line that oilmen must not be permitted to benefit while real citizens suffer from the energy crisis. He went forth to propose a so-called "windfall profits" tax which would strike only on crude prices—leaving the principal operations of major integrated companies unaffected. As bad as the Administration's proposed windfall profits tax is, it is infinitely better than the one which came dangerously close to passing Congress in the final hours before adjournment.

With this backdrop, it is understandable that you and other oil executives would be considering what really is necessary and what is not. But it is shocking that you would have concluded that percentage depletion and intangibles deductions are expendable—and at a moment when our nation most desperately needs to quicken the pace of exploration and development. We must triple our investment in exploration and as soon as possible, if we are to come close to avoiding an energy catastrophy. This we cannot do unless more capital is permitted to flow to those engaged in domestic exploration and development.

To eliminate percentage depletion would do far more to damage domestic development than the figures might show at first glance. Losing the deductible revenue is one thing, but equally important is the fact that depletion and intangibles deductions provide the base upon which the real energy finders attract risk capital into the ventures which develop domestic oil and gas. Surely, a onetime independent oilman such as yourself understands this, even if our critics in Washington and major company executives with no production background do not.

If you are intent on volunteering something to appease the critics, let me suggest that you show real statesmanship and volunteer your foreign depletion and tax credits. This would be entirely logical at this time. There no longer exists any justification for permitting American companies to classify royalty payments to Arab nations as income tax and then deduct this on a dollar-for-dollar basis from taxes owed the U.S. Government. The need is not for more Arab oil development, but rather to bring American international companies home to help independents look for oil and gas within the U.S. We need resource development at home, not abroad. Eliminating foreign tax credits and foreign depletion at this moment in history would be a wholly constructive source.

We are doing neither the consumer nor the nation any favors in yielding ground to industry critics at this moment. Our nation needs a virile domestic energy development industry as never before. Oilmen are no different from other citizens; we aren't going to continue producing oil for less than it costs us to replace it. If we are to be punished by our critics or have essential tax provisions given away by major companies everytime economic conditions begin to permit us to generate desperately needed capital, then independents can be written off permanently. We can never make a comeback. And with us will go all hope of prosperity and perhaps to its very survival. We've too long swayed to the irresponsible voices of a few who have discovered it is good politics in New England and Wisconsin to condemn oilmen no matter the situation. It's high time for the responsible voices in Congress to be heard. I believe the American people can and will understand our plight.

Sincerely yours,

Johnny Mitchell

April 23, 1974

TO: Honorable Henry M. Jackson
Senate Office Building
Washington, D.C. 20510

Dear Senator Jackson:

A long letter to someone as busy and important as you are should begin with an apology for its length.

People not familiar with the energy situation might wonder why a Texas independent oilman would be writing an open letter to an important United States Senator from the non-oil State of Washington. But those of us who try to keep up with these things have long recognized that you have the energy ball more firmly in hand than anybody else in this nation today.

You need no reminder also that you are the author of the provision in law exempting stripper wells from price controls, leaving them in the same category price-wise as foreign oil. Now you are reportedly authoring a so-called "Standby Energy Emergency Authorities Act" through which you would repeal your own stripper well exemption and require the President to establish ceiling prices on all crude oil, including new, released and stripped production.

My contention is that it would be unnecessary and unwise, and very demoralizing, to roll back any crude prices by law. It would be particularly damaging to independent wildcatters, since you are talking about rolling back prices on new and stripper production, where the independent predominates.

The purpose of this letter is not to join the army of oilmen trying to carry on a fuss with you. On the contrary, after looking at what some of your fellow lawmakers are trying to do to the oil industry, I've concluded it's time to recognize that the much-maligned Scoop Jackson is about the best hope we have for coming out of this energy mess with a sensible national policy.

I'm talking particularly about what happened in the House Ways and Means Committee—where members of Congress and their staff tax experts undertook not only to get depletion but also impose a so-called windfall profits tax on us.

By comparison, a crude price rollback is infinitely less damaging to our industry, and especially to domestic independent explorers, than is gutting our percentage depletion provision. Crude prices will ultimately reflect supply and demand, no matter what the politicians do. But juggling depletion, especially if it's phased out entirely as the Committee wants, can do irreparable damage over the long haul. Considering that it's through depletion that really active independent wildcatter attracts risk capital, the net results will be to turn over this industry to a handful of major integrated companies which couldn't find enough oil to keep this nation in lighter fluid. Furthermore, this is one sure way of seeing that the full costs of finding and producing oil is shifted to the consumer in the form of higher produce pricing. Under depletion, a big part of the consumer cost is picked up by investors seeking a fall of the dice through investment in oil ventures.

Just in case your haven't had a chance to analyze fully what the House Ways and Means Committee did to us, let me tell you the way it looks from here.

Chairman Wilbur Mills and his wrecking crew heard a lot of testimony from the industry and the Administration, all pointing up the fact that tax laws had been written to encourage too much oil development abroad and too little at home. So what happened? They came up with a bill which would impose additional taxes during the next six years on domestic production of $13 billion and on foreign production of only $3 billion. What's more, the "windfall profits" tax they propose is a flat excise tax—not on the international companies which have gotten the windfall profits but upon domestic crude. All this adds up to an attempt to get—in a far more damaging way—the crude price rollback that you were seeking in a direct manner. I don't think either course is good legislative policy, but there's absolutely no question but what your proposal would be better for the nation that the silly thing that Mills' Marauders have come up with.

On depletion, the executives of some of the major companies concluded that they were so ashamed of their "obscene" profits that they advocated getting rid of this "albatross around our necks." They weren't speaking for us little independents. We need depletion and intangibles to keep finding oil. Nevertheless, what did the Ways and Means Committee do? It went along with the majors—providing a phase-out of depletion for all of us. The only concession made to us independents was that we could keep a 15% depletion rate a few months longer than the majors.

To prove it was an unusual season at Ways and Means, they reversed themselves to allow a continuation of percentage depletion for natural gas production. Obviously then, the objection was not to depletion as a tax loophole but rather to get depletion as a means of getting the oil industry.

Additional evidence that it was a "get oil" vendetta came when the committee rejected a motion to consider depletion of other minerals along with oil.

Not only did the committee agree to phase out depletion entirely for oil and impose a windfall profits tax on domestic crude, but it even went further to outlaw the existing import fees on imported crude and products.

You have stated publicly that you see a need for continuing percentage depletion for the small independent operator. As a matter of fact, I believe you have stated that, while the term "depletion" is poorly received because it has so long been associated with tax loopholes, you would favor increasing depletion for the true independent up to 35%—and eliminating it for the major integrated companies. I believe this makes good national-policy sense, and I very much hope when this thing hits the Senate Finance Committee, you will make known your views in the most positive way possible.

We independents are taking a bad rap for the excesses of the international oil giants. We are asked to stand hitched for this trimming.

If excessive profits are accruing to oilmen, there are responsible ways to deal with that. And it doesn't include cutting off depletion or rolling back prices on that new and stripper crude belonging mainly to independents. If you have a little room on your campaign special, I know several independents who do give a damn what your company party line may be.

This is no time to shoot from the hip, something I believe the Congress has been doing too often lately. If we don't find answers to the energy problem, there won't be enough left of this nation to justify all the efforts you are making to the Presidency. I happen to be among those who believe that from you we can expect honest leadership out of this mess. My plea is that you will look carefully before taking the next step.

More depends upon what you say and do about energy policy than perhaps you have yet realized.

Respectfully yours,

Johnny Mitchell

June 27, 1975

TO: The Honorable Bob Eckhardt
1741 Longworth House Office Bldg.
Washington, D.C. 20515

Dear Bob:

Let me begin by stating what you already know, that I have confidence in your personal integrity. Moreover, there can be no doubt you are a capable Member of Congress today. A man I do respect even though I don't agree on most issues that you support.

Therefore, I must assume that the oil price measure you successfully pressed through the House Commerce Committee represented your honest evaluation of what is appropriate for our industry and our country. In that spirit, I take issue with you. The facts don't support your position and I want to tell you why.

It seems to me that the Eckhardt scheme for dealing with crude prices is essentially fair only to the large integrated companies. As a matter of fact, I am inclined to believe that you would not have offered this program had certain major company officials from time to time not indicated this program could get along fine on $7.50 crude oil at the wellhead. In that, they are telling you correctly. But they were talking about themselves, and not what's necessary for the industry or the nation.

The issue is not how well existing companies, large or small, will do under any given price level. Rather the issue is what price level is necessary to maximize this country's too-long-delayed search for vital oil and gas resources. In other words, what will it take to get this country moving as rapidly as possible toward the level of exploration and drilling which existed in the mid-50's, when we were last finding enough oil and gas to meet domestic consumer requirements.

I fully realize that Members in Congress would like to believe there are ways to get these results without causing economic pain to consumers. But the sad truth is that we can't and we won't fully develop our energy sources in this country until it makes economic sense to go all out.

Of course you could hide the price realities from consumers if you went the tax route, but Congress just proved it wants the domestic producing industry to pay more not less taxes by curtailing our depletion provision. Now there's a lot of talk about a "windfall profits" tax and even a current hearing before the Ways and Means Committee on "recapturing intangible drilling cost deductions."

Admittedly, to gear up for maximum effort at energy independence would allow profit levels for certain companies higher than generally considered desirable. Even the so-called "windfall profits" is preferable to your ceilings if you provide a full 100 percent plowback goes into increased exploration and drilling.

As one oilman, typical of a good many thousand independent oilmen, who knows how to find oil and gas. It's the wellhead price which determined whether we do or do not. We are unwilling to go all out on a drilling program to find $10 or $11 of oil when politicians are threatening us with a $7.50 limit. Most of the oil we know how to find at today's cost of drilling and equipping wells would be unprofitable at less than $10 per barrel. If consumers and the politicians aren't willing to pay us that much, then they must continue relying on the Arabs.

If you've looked lately at the cost of drilling and equipping a well, you know that a $10 per barrel replacement cost is not excessive. As a onetime oil lawyer yourself, you realize all you have to do to check out whether our end of the business is too profitable is to put some of your own money and your colleagues in Washington into a drilling deal and see how you come out. Chances are you will wind up with only a deduction for intangible drilling costs. But even if you do get production, it will be a long drought before you reap windfall profits at $10 or $11 per barrel. At your $7.50 level, it's not worth trying for.

What disturbs me as much as anything, is that the Eckhardt plan discourages the independents with whom you have long identified. It's the major companies who have their $5.25 "old" crude raised to $7.50 under your plan. Independents, at least the typical wildcatters, don't have any significant "old" oil. The majors have a monopoly on that, and your plan singles this oil out for an adjustment upward—while rolling back new oil by some $4.00 a barrel. Such a rollback would significantly reduce any future exploration and development.

Bob, it takes a pretty good American to admit a mistake in judgement. One that shuts the door to the future energy security of our nation.

I've walked down the halls of Congress these past few years, and I've met a few such great Americans in the Congress. Your sensible colleagues will respect you for facing hard economic facts, those that could encourage the independents to step up their exploration and development of new domestic oil and gas.

Respectfully yours,

Johnny Mitchell

August 14, 1975

TO: Mr. Johnny Mitchell

FROM: The Honorable Bob Eckhardt
The United States House of Representatives
Washington, D.C.

Dear Johnny:

Thank you for your letter commenting on my support for the approach to domestic oil pricing contained in H.R. 7014.

As you probably know, the pricing section of H.R. 7014 underwent some revision when it reached the floor of the House during July.

While the basic approach to oil prices remains the same as in the provisions recommended to the House by the Commerce Committee, the alternations which were made should satisfy some of your concerns about the original bill.

The revised pricing section is designed with the assumption that independent oil producers will continue to receive a phased down depletion allowance which the major producers will not receive. Also, under the new pricing section, "old oil" would continue to be held under the stationary price ceiling of $5.25 per barrel. Therefore, the windfall that might have accrued to the majors if "old oil" had been released to a $7.50 ceiling will not materialize under the reconstituted pricing provisions of H.R. 7014.

An additional concession to independent producers was made by the establishment of a category for genuinely expensive-to-produce oil exempt from the $7.50 average price level that would govern most domestic "new oil." An independent producer whose drilling and production costs justify $12 or $14 per barrel may get that price so long as the average price of crude in the category does not exceed the price of $10 per barrel. The $10 figure is an average price ceiling for all oil produced in fields certified as requiring exceptionally expensive recovery efforts rather than an inflexible limit on the height to which domestic oil prices may rise. It is my belief that this category will apply primarily to independent producers.

I am convinced that a pricing system of this sort is necessary if the U.S. is to develop sufficient supplies of domestically produced oil and at the same time take the pricing discretion over domestic oil out of the hands of the OPEC cartel. If domestic oil prices were set at $15 per barrel we might profitably produce more oil than we would at the figures set in H.R. 7014, but would the incremental increase in production justify the impact such a price escalation would have on the American economy?

The experience of two years of energy price inflation and the related recession suggests to me that such an impact cannot be justified. While energy needs are certainly a factor in calculating the right price for domestic oil, we must also consider the needs of a healthy national economy. For the sake of economic stability and in the interests of a successful recovery from the disastrous recession of the past few months, I believe we must hold oil prices to their present plateau for the immediate future. I am convinced that H.R. 7014 accomplished this objective.

I appreciate this opportunity to have your views and to share mine with you. Please do not hesitate to write whenever I can be of assistance.

Sincerely,

Bob Eckhardt

December 5, 1975

TO: The Honorable Frank Zarb
Administrator
Federal Energy Administration
Washington, D.C. 20461

Dear Mr. Zarb:

As an independent oil producer who has held you in high esteem I must express deep disappointment in the course you have taken with regard to the horrible omnibus energy measure.

Upon realizing you had been trapped into agreeing with the maneuver by Senator Jackson and his cohorts, instead of acknowledging you made a serious blunder and so advising the President, you've sought to protect your own flank. By attempting to justify your course in political terms, implying that this is necessary to the reelection of President Ford, you've asked the President to compound the error by ignoring his other advisors and joining you in approving the bill.

According to my information, you have stated that the only alternative to his signing S-622 into law is total and instant de-con-trol of oil prices. It is understandable that Bo Calloway and those whose primary role in politics would crawl in bed with you on being given this kind of information, but you know far better than most that there is another course.

If the President vetoes the measure, there is absolutely no reason why the present control system cannot be extended for another fourteen months are so, during which time reasonable measures can be pursued to phase out controls on old oil. This would be a course the

President could live with, and could safely sally fourth into the New Hampshire and other primaries without damage to his political cause. That his political enemies don't want to see this constructed course pursued is no reason for you to reject it.

Also according to my information you've personally talked Republican Governors in their Wichita, Kansas meeting out of adopting a resolution urging veto of S-622. You recordedly implied such nonsensical arguments as that the measure resulting from the Jackson-Staggers Conference Committee had not yet reached the print out stage, and hence the judgement could not be made on the merits of the measure. You knew full well what the "conceptual agreement" consisted of and you must surely know it would bring disaster to our industry and our nation. The domestic oil industry can't live with this measure, and if you've understood and agreed with anything you yourself have been saying in recent months you surely understand and agree with that.

Electing President Ford to a full four-year term is important to the nation, I happen to believe. But it is not as important as moving this nation toward the goal of energy independents as early as possible. You do the President no favors by assuming he is more concerned with his own political future than with the well-being of the nation.

If you prevail, and if the President signs this terrible bill, you and he must share a great responsibility for undoing all the good you've done in the past. Up to this point you've stood firmly against the demagogues led by Senator Jackson. Now suddenly you're letting them call the tune, and implying that you have no alternative. There is a reasonable way out of this, consisting in the extension of the present arrangements for the time being—and I don't believe doing what's right will prove to be bad politics in the end. On this one you're dead wrong, Frank, and if you are the man I think you are, you'll stop compounding your error. By urging the President to join you in your mistakes is a course unworthy of a fine public servant such as yourself.

I think I would be remiss in my duty if I did not tell you I sincerely believe that should this bill become law, it'll take a very short time to realize how disastrous it is. Unless the government wants to put oilmen in uniform, I don't believe they will continue to produce their oil at less than replacement costs. We did exactly that for a good many years hoping for an awakening as to what was happening. Now that the realization has come and President Ford has undertaken to do something about it, we do not propose to let it happen again. Not only will we not carry out our drilling program, but even current production will be cut sharply if the President goes along with the outrageous maneuver by Senator Jackson and his cohorts.

Respectfully yours,

Johnny Mitchell

March 9, 1976

TO: The Honorable Bob Eckhardt
1741 Longworth House Office Building
Washington, D.C. 20515

Dear Bob:

As one who insists on regarding you as a friend, one who is convinced you honestly believe yourself to be concerned with the well-being of the independent oilman (even while most independents seem increasingly to disagree), I take the liberty of offering additional comment on recent developments.

The charge has been made, one you deeply resent, that you have lost sight of the well-being of an important industry in your State because of your zeal to claim a leadership position with those liberals in Congress who have chosen to make the petroleum industry their favorite whipping boy. I have followed your record from the earliest days in the Texas House of Representatives, and I agree with you that you have followed a pretty consistent course throughout. Except in certain limited respects, it seems to me the causes you are espousing and the actions you are taking today are pretty much in line with those of the past.

One exception which comes to mind pertains to your carrying the natural gas pipeline flag on the subject of expired contracts.

Back when the Texas State Legislature attempted to assert limited jurisdiction over natural gas prices—in order to end the rankest kind of discrimination against small producers, royalty owners, and the

State of Texas—the pipeline purchasing companies fought us bitterly. Their battle cry then was "sanctity of contracts," you will remember, saying that where they had contracted for gas for as little as 2¢ per thousand cubic feet no governmental body should be allowed to interfere with the "Sanctity" of those contracts. They won that battle in Texas—with the result that many got trapped under federal regulation at absurdly-low prices.

Now the question before Congress is whether gas under contracts which expire should by law be locked into those purchasers forever at regulated prices. No longer do we hear the gas pipeline companies defending the principle of "sanctity of contracts." Instead, they maintain that to protect the public they must be allowed to continue taking that gas at the old regulated price, even though the contracts under which they originally bought the gas no longer exist. This is not surprising. The surprise to me was to hear you take up their battle cry on the floor of the House of Representatives during the recent debate on gas deregulation.

My more fundamental difference with you goes to the question of price controls on both oil and gas.

You've many times registered quite effectively your belief that OPEC should not be allowed to dictate crude prices here, which you say would be the effect if crude price controls are eliminated. I say that if price controls really are necessary, they should be at the consumer or marketing level—with ripple effects back to the wellhead. This would protect the consumer interest, to the extent he is protected by any kind of price controls, while leaving free that segment of the industry which most desperately needs encouraging to make us independent of OPEC. Yet the Congress mandated in EPCA, to which you contributed mightily, that we decontrol at the downstream level but leave wellhead price controls intact—and even rolling back the price of new and stripper crude oil.

In principle nobody can differ with you that a foreign cartle should not be allowed to dictate prices here. But I'm not sure that wouldn't be better for America than letting politicians who aren't interested in determining what is necessary to get the wells drilled dictate prices. Impartial studies, such as the one by Bob Nathan, a liberal economist with whom you are personally identified, have concluded that it now costs about $12 per barrel to develop new oil in this country. Yet you and other politicians have mandated that we can get only $7.66

average. Maybe you think it's not important that America emerge independent of OPEC. Maybe you think we don't need to develop alternative fuels in this country, which will require the equivalent of at least the $12 per barrel figure.

On natural gas, I think you're even further off base. You just can't be unaware that the average price for gas in this country is far below that of less-desirable competitive fuels. You know, I feel certain, that most independent producers have not historically even looked for natural gas, because there has been no incentive. We found gas because we looked for oil. We don't even know how much gas we could develop if the nation's 10,000-plus independents were really turned loose to concentrate on it. We got a glimpse of what can happen when the intrastate market was recently allowed to move up to about $2 per thousand cubic feet—with the result that we quickly developed a surplus where a serious shortage was already being realized.

Bob, what makes you the big fly in our ointment is that you are recognized in Congress and in our industry as an extraordinary man who knows something about the oil business. You practiced some oil law, and are a brilliant man by any standard. When you stand forthrightly against allowing wellhead prices of both oil and gas to arrive at replacement-cost levels, many a good man from so-called consuming states thinks the industry must be lying about its needs.

You're not the only person worried about inflation. Everybody's concerned about inflation, but the way to stop it is not to freeze the price of those commodities whose supply we need to increase. The way to stop it is to head off some of the unnecessary increases in commodities which are in surplus supply, like automobiles. I don't see you doing a hell of a lot about those other commodities. You seem to concentrate on the leading industry of your own State, to the serious detriment in my opinion of that industry and your own State.

I suggest you're listening to people who aren't really on your side, people who have a distorted view of what a good Congressman should do.

I sincerely and urgently request that you reconsider your position in these matters.

Respectfully yours,

Johnny Mitchell

May 10, 1978

TO: Mr. Johnny Mitchell

FROM: W.E. Turner
Executive Vice President
TIPRO
Austin, Texas

Dear Johnny:

I'm pleased that Gulf Publishing Company has commissioned you to publish your communications over the past two decades. The letters you exchanged with all and sundry do indeed provide an interesting and unique perspective on the events transpiring in our industry during a period when domestic wildcatters occupied last place on Washington lists of national priorities.

What is today properly called "America's energy crisis" could have been averted. We could today, and for the foreseeable future, be a nation self-sufficient in energy. Had America pursued the energy policy you urged, we could today be secure in an insecure world, enjoying a favorable balance of payments and a favorable balance in our economic and military posture. We are instead a nation in jeopardy. And as America goes, so goes the free world.

When you screamed out warnings so long and so loud, many misunderstood. Some admonished you to sit down and shut up. Your closest friends and advisers, even your family, were often implored to urge you to cease and desist. Too few of us stood with you throughout, even though we realized you were fundamentally on course. Your own brother George, a true business genius immersed in his battles to build an empire, knowing his own limited influence over your actions, developed a stock answer: "You want a brother, I'll give you one."

Those of us who knew you best never questioned your intuitive wisdom or your unerring inner direction. We had learned you have an uncanny instinct for viewing things in proper perspective. Our concern was that you might choose the wrong words to convey your message—and thus discredit yourself and your very important message.

If as you say, Jim Clark and I continually harangued against some of your letters and speeches, cautioning you to temper your words, it was not out of any conviction that you were off course—and certainly not because we felt what you were doing was unnecessary. I fully shared your outright alarm at our nation's headlong fall into helpless dependence upon clearly unreliable sources for the lifeblood of our economy.

When even a President of the United States (under whose administration, incidentally, domestic exploration and drilling suffered an unprecedented decline) got so upset with you as to suggest strongly to mutual friends that you be called off, I greatly admired your courage in standing firm odds. A lot of us, including Jim Clark and myself, may have scurried for cover, but we never really lost faith in the soundness of your mission.

Maybe like Don Quixote, you only tilted at windmills, but they were the windmills which needed tilting.

It must seem all the more gratifying now to witness major companies, the same major international companies you implored to turn their emphasis homeward, arguing your case for lessening America's dependence on foreign energy sources. And it must seem strange to hear a President, who is personally no friend of the oil industry, saying as you did for two decades that we must restore incentives necessary to the rebuilding of America's domestic energy might, at whatever discomfort to consumers who always fancy their interest lies on the side of cheaper prices.

In short, Johnny, I was proud of you throughout and I'm proud of you now. However small and insignificant in the eyes of some, you did play a role in a period of crisis for our industry and our nation. You were fundamentally right throughout, even if your methods may have left some of us quaking in our boots. While you never exactly occupied center stage, you became a force to contend with, managing somehow to get the attention of those high in government and industry who would have wanted nothing more than to pretend you didn't exist.

In these letters, at least, is evidence that you unlike so many others in our industry and government are no Johnny-come-lately to the concept that our well-being as an industry and as a nation requires an energy policy which makes it possible to develop our own ample energy resources. We've now learned the hard way the cost of doing otherwise. Your distinction is that you saw it all clearly before we had to learn that painful lesson. You are entitled, through publishing these letters, to say in good conscience "I told you so."

Warmest personal regards.

Most sincerely,

W.E. Turner